The Book of Nature

Northeastern University 1898–1998

The Book of Nature

Natural History
in the United States
1825–1875

Margaret Welch

NORTHEASTERN UNIVERSITY PRESS
Boston

Northeastern University Press

Library of Congress Cataloging-in-Publication Data
Welch, Margaret, 1961–
The book of nature : natural history in the United States,
1825–1875 / Margaret Welch.
p. cm.
Includes bibliographical references and index.
ISBN 1-55553-342-6 (cl : alk. paper)
1. Natural history—United States—History—19th century.
2. Natural history literature—United States—History—19th century.
3. Books and reading—United States—History—19th century.
I. Title.
QH21.U5W45 1998
508.73'09034—dc21   97-43421

Designed by Gary Gore

Composed in New Baskerville by Coghill Composition Co., Richmond,
Virginia. Printed and bound by Thomson-Shore, Inc., Dexter, Michigan.
The paper is Glatfelter Supple Opaque Recycled, an acid-free sheet.

MANUFACTURED IN THE UNITED STATES OF AMERICA
02 01 00 99 98   5 4 3 2 1

In memory of
C. M. W.
and
R. E. W. Jr.

With love and respect

Contents

Illustrations

Acknowledgments

This project began as a senior honors thesis at Williams College and evolved into a dissertation for the American Civilization Department at the University of Pennsylvania. Thanks in large part to a research fellowship from the Advanced Studies Program, Winterthur Museum and Library, I was able to move into different areas and to develop new themes. I therefore owe thanks to too many individuals and institutions over the years to mention all, so only my most recent debts will be named. However, I keenly remember studying the original Audubon works as an undergraduate and thank the staff of the Chapin Library at Williams College for the privilege. I also wish to thank my dissertation advisor, Karin Calvert, and my other readers, Krista Wilmanns Wells and the late Anthony N. B. Garvan.

The staffs and collections at the following institutions provided most of the material for this effort: Academy of Natural Sciences (Philadelphia), the Ewell Sale Stewart Library; American Museum of Natural History (New York), Library; Johns Hopkins University (Baltimore), George Peabody Library and Milton S. Eisenhower Library; Library of Congress (Washington, D.C.), Rare Book Division; Natural History Museum (London), Library; Princeton University Library, Department of Rare Books and Special Collections; Smithsonian Institution Archives (Washington, D.C.); Yale University, Beinecke Rare Book and Manuscript Library. I wish to thank the Smithsonian Institution Archives and the Beinecke Library for allowing me to reproduce textual material. I acknowledge also the institutions who permitted me to reproduce illustrations and the many helpful individuals who aided me in the process. Herb Crossan graciously offered his photographic work.

The persons who agreed to read the manuscript—Allan Holtzmann, Bert Denker, and Neville Thompson—provided guidance and the most tactful criticism. Allan, in particular, provided a wealth of information. It gives me special pleasure to thank Bert and Neville, now my supervisor and colleague, respectively, at the Winterthur Library. Pat Elliott, as Winterthur's Advanced Studies Coordinator, made for a pleasantly productive stay during my Winterthur research fellowship. Ron Tyler and Shirley Wadja rendered professional favors as well. My former professors at the College of Library and Information Services, University of Maryland, College Park, graciously permitted my personal research to enter my term papers.

My family, including my siblings, aunts, uncles, and godparents, have never doubted the project. Two individuals deserve special tribute. Jeremy Cott has provided crucial support, and his generosity heartened me during the often difficult process of preparing the manuscript. Elizabeth M. Welch has guided many books through publication, but my rough drafts may have challenged her professional prowess. Both her editorial skill and sisterly love inspired me.

Introduction

The Primacy of the Book

On November 25, 1845, the post-master of Greensboro, Georgia, W. L. Strain, wrote to the foremost American ornithologist of the era, John James Audubon, to "inform [him] that I have in my possession now in this place a fine full grown 'Golden Eagle,' " the catch of a neighboring slave who first believed it to be a wild turkey. Thanks to the loan of Audubon's *Ornithological Biography* (the text accompanying the famed *Birds of America*) from former Congressman W. C. Dawson, Strain was able to compare his specimen to its portrayal and happily assured Audubon that "if you had had the one in my possession before you, you could not have described it more minutely and accurately than you have." He himself gave its physical dimensions much as Audubon had described his specimen: "[It] measures 6 feet 11 inches from the tip of one wing to the tip of the other, is about 3 feet from the end of the beak to the top of the tail, and I should judge weighs 12 pounds." Strain invited Audubon to see "the noble bird" if he was ever nearby, as Strain chose not to send it to him, preferring instead "to keep it for all to look upon a fine specimen of the Bird Creation."[3]

Roughly two decades later, a trio of Cambridge, Massachusetts, boys were using the *Ornithological Biography* to guide their bird-watching and egg-collecting. One of these young ornithological devotees, Daniel Chester French, amusingly recalled in later life: "I think I can fairly claim that he [William Brewster] and Dick Dana and I made a more serious study of the subject than was usual, perhaps because Will's father had Audubon's *Ornithology* and my father had Nuttall's [*Manual of Ornithology*], which we studied with a thoroughness which would have put us at the head of our classes if applied to our school books." French and Richard Dana dropped their birding pursuits and proceeded to their distinguished careers as sculptor and man of letters respectively, while William Brewster's passion only grew. In 1871 Brewster and another coterie of youths formed what would become the Nuttall Ornithological Club, at first "meeting regularly, one evening a week to read 'Audubon' aloud, and to discuss it in the light of their own experiences." Brewster's entire adult life would revolve around bird study, while other club members such as Henry Wetherbee Henshaw also became ornithologists.[4]

At the same time but far removed from a companionable atmosphere like that of the Nuttall Club, the young John Henry Comstock pursued his scientific interests alone. According to the memoirs of his wife, Anna Botsford Comstock, he studied the plants of the Great Lakes region while acting as a ship's cook to earn college money. On a stop in Buffalo he wished to purchase a book devoted to cryptogamous (nonflowering) plants, and the bookstore clerk pointed him to the shelf "with all the science books we have." Comstock found not cryptogams but entomology, in the form of Thaddeus W. Harris's revised 1862 *Insects Injurious to Vegetation,* with its ample illustrations. Comstock received an advance from his captain for the expensive purchase and "learned the orders of insects with the book propped up before him while he was washing dishes." He returned to his studies at Cornell University the next fall and eventually taught entomology at that institution. His notation on the book's flyleaf (an epigraph to this chapter) commemorates its significance in his life.[5]

These three examples underscore the overwhelming importance of books in the transmission of natural history practice and discourse before 1875 in the United States. Natural history books, especially those with illustrations, taught individuals the basic principles and accompanying terminology of the sciences. Comstock memorizing the orders of insects is a case in point. They furnished identities and information about the nonhuman world so that observers could order and feel familiar with the new or confusing flora and fauna around them.[6] Thus William Strain, for example, was able to identify his golden eagle. Americans, whether emigrating to the United States or moving westward, encountered a far greater number of creatures than they could identify, thereby magnifying the role of natural history to explicate the unknown. For those inhabitants familiar with natural history, knowledge of the surrounding environs created a richer attachment to their new home, which deepened with each change of season and new habit or specimen observed.

The young Nuttall Club members were able to compare their various bird-watching experiences with those of Audubon thanks to the medium of the printed book. The language and imagery, once absorbed, could be shared between individuals in conversation or in letters. Books introduced these young men and other Americans of the time to the various natural history disciplines (in this case ornithology), which in turn provided a focus and enrichment to their lives. The hours spent rereading their texts, collecting and drawing specimens, and associating with fellow devotees gave not only emotional satisfaction but a sense of tangible accomplishment. For John Henry Comstock, entomology transformed his life.

Such "intensive" reading of texts and viewing of illustrations passed on the distinctive language and imagery of natural history.[7] The early writings of the Nuttall group resemble the texts of Audubon, Thomas Nuttall, and their illustrious predecessor Alexander Wilson; even Strain, a comparative novice to the discipline, unconsciously mimics the listing of physical attributes common to natural history descriptions in his account of his eagle. Artists like Audubon himself studied natural history illustrations to learn the genre, thus transmitting its visual conventions. The scarcity and expense of the texts only increased their importance to their readers: Comstock protected his copy "from the pervasive coal dust" of the barge, and persons shared and discussed the rare *Ornithological Biography* in locales as diverse as Greensboro and Cambridge. Today's scientists undeniably absorb the information and principles of their disciplines from books, but the mediation of formal education provides a different experience.

Perhaps as interesting and important as these individuals dedicated to their subject are those non-naturalists who employed the natural history genre for specific purposes at various times of their lifes. The explosion of antebellum print culture made the text and images of the scientific monographs available to a large audience, albeit in modified forms. Young women incorporated natural history drawings into their keepsakes, and writers could assume that their readers not only knew natural history but could appreciate satire mocking the naturalist's perspective. Those creating artwork for the home knew and used zoological designs for their own creations. Newcomers to a region wrote about the flora and fauna in the pattern and diction already familiar to them. Surprisingly, historians of science and social historians have bypassed this popularization of natural history, but it was a vital current in the United States in the mid-nineteenth century.[8]

This work seeks to explore the interactions between the natural history books—mainly the major illustrated monographs—and the readers and viewers in the American public between the years 1825 and 1875. It does not focus upon natural history societies or academic institutions' natural history curricula, as both entities were relatively weak in antebellum America and grew in importance only toward the end of the period. Most local societies were short-lived, whereas the major institutions of the later nineteenth century, like the Smithsonian Institution and the Academy of Natural Sciences in Philadelphia, were in their infancy or at low points. Academies and colleges began to include botany but only as an elective, and zoology as a separate subject was not included in most curricula. Those making natural history images were compelled to copy book illustrations, since no schools or private individuals taught the format. Individuals outside the classroom read and looked at and, in many cases, financially supported the original scientific monographs of Audubon and Alexander Wilson, the lesser-known botanical works of Asa Gray and the conchologists, and the United States government expedition reports. Textbook compilers, periodical editors, and other authors copied and discussed these texts through the rapidly growing magazine and book trade. It seems, then, that a social history of natural history in the United States should favor individual response and dissemination over institutional history (the latter strategy employed in David Elliston Allen's landmark study, *The Naturalist in Britain*).[9]

Natural History in the United States

The exposure of individuals like Comstock and groups such as the Nuttall Club to natural history is supremely important because it is as "learned" and highly artificial a discourse as any in Western culture. Natural objects in this discourse historically have not been portrayed as part of everyday experience, as the plants and animals are visually isolated from their environment and verbally described in erudite language. The genre of illustrated natural history books with interlocking text and images, in particular, remains remarkably stable from Greek manuscript herbals to today's wildflower identification books.[10]

Within the tradition, transformations did occur specific to the United States in this period. One prime addition to the genre occupies this book, namely the development of extended "life histories," virtual biographies containing every scrap of available evidence on appearance, reproduction, communication, habitat, life stages, and geographical distribution of a species. The life histories' visual equivalent is the representative specimen(s) in a characteristic pose, often against a generalized background suggesting its habitat. Zoological illustration historian Ann Shelby Blum has established this iconography of the "living animal in nature" as an important American contribution to the existing tradition at the time. Because the dearth of libraries and study collections comparable to those of Europe forced American naturalists to concentrate on their inherent advantage of firsthand observation of species, they consciously emphasized their experiences and knowledge of the species' environs and habits in prose and picture. The rapidly expanding United States, with its tremendous climatic variations, similarly spurred the Americans to contribute insights into the linkage between geographical range and the creation of species.[11]

Historians of nineteenth-century culture are increasingly aware of the cultural ideologies and concerns inherent in a dominant imagery or viewpoint, such as the contemporary panoramic landscape signaling a historically specific relationship with the American wilderness. The emergence in this period of the consistent imagery of the representative specimen against its suggested haunt deserves new attention. This focus on a floral or faunal specimen in text and imagery gives the species the attention usually accorded humans. Tellingly, contemporary naturalists and critics used the terms *biographies* and *portraits* for the life histories and illustrations. This treatment held important consequences for Americans' relationship with this portion of the nonhuman world. It implicitly granted species an autonomy not offered by the traditional anthropocentric view. Moreover, this era explicitly likened plants and animals to humans. Authors and their readers eagerly cast animals and birds as moral analogues and enjoyed linking human qualities to plants through the language of flowers.[12]

From the birth of the new republic onward, those responsible for creating the written, spoken, and visual components of a national culture advocated a distinctively "American" mode of expression and content. As decades of Americanist scholarship have demonstrated, the efforts to create this unique national culture frequently blended European forms and American content. James Fenimore Cooper's adventure novels star Native Americans; mid-nineteenth-century landscapes of native wilderness scenery are ultimately derived from the European topographical tradition; and genre paintings, perhaps influenced by the earlier Dutch images, feature quintessential American characters. This perennial challenge of cultural nationalism stirred the American naturalists in these decades to create remarkably ambitious scientific illustrated books devoted to U.S. species. Not only their avowed intentions but the very substance and structure of the works testify to their desire to match the Europeans through the adaptation of an imported vehicle, the natural history monograph, to manifestly American subjects, the land's flora and fauna.

John James Audubon has long been appreciated as an American "original," but other naturalists, such as botanist Asa Gray and herpetologist John Edwards Holbrook, set the aim of describing and illustrating definitively the species covered in their disciplines within the country's confines. Subsidized by the federal government for many decades, the massive series of survey reports on the newly expanding western possessions brought as an added benefit an unparalleled opportunity to illustrate and catalog more native flora and fauna. Historians of science have long appreciated that Americans relied upon this use of their local environs, flora, and fauna to give them presence in the European community; perhaps Americanists in general, exposed to those famous examples of authors and artists such as Cooper and Thomas Cole, who strove to create original works with imported genres (and a sometimes unsure audience) in the antebellum era, may be interested to rediscover how a contemporary subset of practitioners in the arts and sciences were reaching toward a similar goal.

Grandiose national programs were evident in these monographs and surveys intended to portray literally the nation's products in a manner unsurpassed internationally. However, these works encompassing the entire country were ultimately based on individual observations often born of love for a particular locale and its natural occupants. Resident observers, often as dedicated to the natural history discipline as such famed authors as Audubon and Asa Gray, also communicated their findings on the occurrence and habits of species. This information, gained through years of sightings and note taking, was made known through either publication of more modest local lists or direct communication with the major authors; these local observations provided much of the substance of the national works. Thus specific and localized efforts balanced the grandly nationalistic ambitions.[13]

Art historians have provocatively raised the issue of sectionalism versus national unity in pre–Civil War landscape images previously believed to be apolitical; artists and cultural leaders of the Northeast, it is contended, intended to establish their countryside's scenes as the national ideal.[14] The promotion of natural history by naturalists and those editors publishing the excerpts from the scientific monographs may also be interpreted in light of the growing sectional conflict. The imperative of the national monograph required naturalists to study all species, so northern and southern naturalists chose to ignore political differences in their collaborations. Publishers and editors may have emphasized (on a conscious or unconscious level) natural history, with its stimulating and diverting content, to turn attention from such troublesome issues pertaining to the human species as states' rights and slavery.

Creation and Response

This book's two-part organization reveals its twin interest in the creation of and reaction to the natural history works. The first chapter describes the tradition inherited by the antebellum generation, in particular the genre of illustrated monographs devoted to a portion of U.S. flora and fauna as epitomized by Alexander Wilson's *American Ornithology*. Chapter 2, "Creating the Inhabitant Image," discusses how the well-documented example of Audubon's ornithology as well as the lesser-known yet similarly significant botanical and zoological monographs established the important life history format, with its corresponding iconography. Chapter 3, "Nature Writ Large and Small," relates how state and federal governments enabled the continuation of the illustrated scientific monograph tradition through the publication of survey reports while private individuals persevered in developing their own contributions. Both chapters 2 and 3 illuminate the crucial role of fellow natural observers, often isolated individuals or groups scattered throughout the nation, in donating specimens and personal accounts to the national works.

The burgeoning appreciation in historical readership and historical viewing shapes the second part's focus upon dissemination through print culture and individual responses.[15] Textbooks imparted empirical knowledge derived from natural history to readers, many of whom were outside the classroom. Other important literary genres such as children's literature, sentimental flower books, and biblical histories borrowed natural history for their own purposes. Non-naturalists read closely and incorporated the highly distinctive natural history rhetoric to respond to specific incidents throughout their lives. Travelers' diaries, for instance, relating new creatures in quasi-scientific prose, are described in chapter 4. The following chapter, on the response to the visual imagery, uses the example of Audubon's subscribers as a group to examine motivations for patronage of natural history. The unprecedented increase in available natural history imagery inspired specific adaptations in mass-produced artifacts and handmade creations for the home. The conclusion traces the persistence of the "living animal in nature" and the life history tradition to today's nature films, photography, and conservation legislation.

The transmission of natural history texts depends on literacy. Therefore, this study cannot incorporate the rich interpretations of their natural environs by many nineteenth-century Americans, such as Native Americans and African Americans, who generally did not participate in print culture and were perceived to hold values antithetical to natural history. Other disciplines popular in the era, like geology and astronomy, are not covered, as the legacies of those disciplines later to become known as the life sciences appear of more lasting significance to today's popular culture.[16]

Historians have consistently aligned this period with the widespread passion for the vague yet frequently invoked "Nature," from Perry Miller's lofty analysis of America as "Nature's Nation" to E. Douglas Branch's amused characterization of U.S. citizens as sentimental "nature addicts."[17] To this author, there were (and continue to be) two groups consistent in U.S. history, those who desired and were able to identify "a hawk from a handsaw" (to use a favorite nineteenth-century phrase) and those who could not and would not. The former were a comparatively minute band of dedicated amateurs and professionals interested in systematically identifying, collecting, and often publishing discrete segments of the natural world. Today's birders who pursue "lifelists" are easily recognizable heirs to the tradition. Exhaustive reading and viewing of natural history texts, whether in childhood (as in the case of William Brewster reading the ornithologies more often than his assigned schoolwork) or in adulthood, frequently triggered and sustained these individuals' interest in the specific subject. They channeled their intellectual curiosity and aesthetic sensibility into a lifelong avocation.

Those apart from these nature observers and naturalists represented the bulk of the population. They were not actively seeking out, collecting, and classifying birds, snails, or wildflowers throughout the years. As in Daniel Chester French's case, the youthful energy was channeled into another passion. However, because this second group was highly receptive to "Nature" and the natural world in the mid-nineteenth century, men and women of the upper and middle classes in most parts of the country were reacting to natural history's descriptions and imagery in a variety of fascinating ways. Evidence including hunting manuals, nonacademic paintings, commonplace books, and marginal notes on the books themselves suggests that Americans across geographic and gender boundaries were using the imagery and language of the naturalists not necessarily to pursue natural history but to express their individualized relationships to the natural world.[18]

This study evolved from a dissertation on the relationship of Audubon to his patrons to include the rich legacy of other naturalists, nature observers, texts, and imagery. Other contemporaries pursued natural history interests with an ambition and vigor equal to Audubon's: from John Godman, one of the first Americans to write a book on native mammals, who "would walk miles to observe a shrew," to Annie Law, who climbed the mountains of Tennessee to obtain land snails. Monographs such as William Henry Edwards's beautiful *Butterflies of North America* and the homegrown *Illustrations of the Nests and Eggs of Birds of Ohio* by the Jones family are only now receiving the attention due them. Individuals not commonly associated with natural history, such as ethnologist Lewis Henry Morgan, the "Brandywine sisterhood" members Eleuthera and Sophia du Pont, Maine parson-painter Jonathan Fisher, and little-known or anonymous diarists, government survey artists, and craftpersons, devoted hours of intense effort composing their commentary on the natural world. Their efforts also deserve recognition.

1

The Tradition
before 1825

> ❧ *Much as we are indebted to the efforts of the natu-*
> *ralist, to whatever nation he may belong, it is yet a*
> *mortifying truth to our literary pride, that by foreigners*
> *alone, has not only this, but almost every other branch*
> *of our natural history been illustrated.*

Alexander Wilson in an 1807 prospectus for his
American Ornithology (1808–1814)[1]

*A*lexander Wilson (1766–1813)
thus acknowledged, albeit reluctantly, the firm base of depiction and
description of New World flora and fauna that he and other Americans inherited. Natural history, particularly in the form of illustrated
books, had developed a consistent and distinctive imagery and language by the time this poet-naturalist adopted the discourse in the
beginning of the nineteenth century. Earlier European travelers employed the mode of description in order to express their experience
of the New World's plants and animals. Despite its seeming rigidity,
the tradition could incorporate these shifting subjects and individual
creators so well that Wilson and his contemporaries adapted it for
their purposes. The succeeding generation took over the enhanced
tradition.

Wilson's admission of the "mortifying truth" signifies Americans' dilemma related to Europeans' extensive study of New World species. Many species had been and in all probability would continue to be discussed in foreign works of intellectual substance and physical beauty, so American natural history publications may have appeared unnecessary. Vigorously asserting the need for Americans to publish their own works, Wilson articulated explicitly in his promotional literature, introductions, and even the species descriptions his position that advantages springing from residency in the United States uniquely qualified its citizens for natural history duties. *American Ornithology*'s innovations, including the substantial life histories devoted to each bird, suggest (perhaps more convincingly than Wilson's pronouncements) that the opportunity for increased study and more intimate knowledge of habits led to more thorough and accurate portrayals. Future naturalists would echo Wilson's claims of American superiority in regard to observational powers and would continue to exploit the life history format he championed and to enhance the iconography he suggested.

American naturalists, like other authors in the early republic, worked in an atmosphere that craved cultural achievement yet found it difficult to promote. Because the educational, governmental, and corporate structures that support twentieth-century scientific endeavors had not yet coalesced, private individuals were required to fund natural history publications. In a still undeveloped national economy, those blessed with surplus wealth often preferred to spend it on items of obvious personal value, such as portraiture and fine home furnishings. Disinterested sponsorship of native artists experimenting with history and genre paintings or literary figures writing novels or poetry was difficult to muster, despite the frequent calls by nascent art societies and magazines for the fostering of an American identity. Wilson would have to defend his project in a clime generally perceived as inhospitable to nonutilitarian endeavors including natural history. Wilson's success in producing an unprecedented project encouraged later naturalists, especially John James Audubon, likewise to attempt ambitious illustrated monographs devoted to U.S. species.[2]

The Natural History Genre

In works preceding Wilson's *American Ornithology*—most frequently in travelers' literature—natural history discourse played a major role in defining the land that would become the United States. Descriptions in the natural history mode often appear in the accounts of Europeans who visited or settled in the New World. The persis-

tence of these natural history inserts in travelers' personal narratives
may surprise the modern reader, yet the flora and fauna indigenous
to a region was thought to characterize it as much as its other "natural productions," like human inhabitants and climate. (In a related
tradition, until the mid-eighteenth century maps often included animal portraits in the margins or on the land itself; species that seemed
exotic to Europeans, such as the armadillo and cacti, decorated car-
tography of the New World.) Most of the early European exploration
texts, such as those of John Smith and Francesco Hernandez, list species and their habits when observed by or known from others. European settlement brought forth more descriptions of the new land's
inhabitants, as in Thomas Morton's *New English Canaan* (1637), with
its entries on twenty mammal species, thirty-two birds, twenty fish,
and twenty-seven plants. William Wood in *New England Prospects*
(1634) enlivened the standard prose list by placing his description in
verse ("Quill darting Porcupines, and Rackoones bee / Castell'd in
the hollow of an aged Tree"). Eighteenth-century American history
classics such as John Lawson's *New Voyage to Carolina* and William
Byrd's *History of the Dividing Line* contain not-so-famous animal and
plant lists. These botanies and zoologies "of place" would continue
in the form of state and local histories into the nineteenth century.[3]

The advent of printed illustrated natural history books coincided
roughly with the European Age of Discovery, and the development
of woodcut blocks allowed for illustrations' integration with texts. A
few New World species, like the turkey and porcupine, entered into
the sixteenth-century encyclopedias of Conrad Gesner and others,
which remained influential in conveying zoological information well
into the next century. The herbal format—the compilation relating
primarily to medicinal and food plants, embodying most of the botanical information produced since ancient times—similarly showed the
new forms of American maize, potato, and tobacco. Challenged by
new natural forms, European illustrators worked from dried specimens or written descriptions. Their results may appear awkward to
today's viewer, but, because these were the best images available, they
were copied widely not only in other encyclopedias but in other animal picturing. The herbal artists were also challenged to draw anew
rather than rely on the older forms sometimes derived from early
Greek herbals, as exemplified in Illustration 1.1, Thomas Johnson's
depiction of the exotic banana received from Bermuda, described in
his 1633 edition of John Gerard's *Herball*.[4]

Natural history prose and iconography are so distinctive that one
may easily distinguish the discourse in the travelers' accounts and

encyclopedias despite their ranges of date and nationality. The species of plant or animal is conceived as one entity with consistent characteristics, not as a group with individual differences. The divisions within this species are based on obvious physical attributes like age and sex (plants—blossom, fruit; animals—male, female, young). The habits, climate, and general haunts (e.g., swamps, mountains) are assumed to apply to the entire species whether depicted in the image (as in a night scene with an owl) or described verbally. Individuals within the species are assumed to share these characteristics with other members, which in turn distinguish them from other entities. The discourse explains every physical attribute and intends to eliminate ambiguity for the reader/viewer. For example, sixteenth-century encyclopedias stated both physical attributes and qualities now regarded as fable as undoubted, demonstrable truth—unlike contemporary emblematic images involving the same species, which required considerable prior knowledge on the part of the reader/viewer to decipher the intended message. The imagery presents as clearly as possible without obstructions or vicissitudes of lighting and atmosphere the ideal specimen. The lack of light and shade lends a highly unnatural, atemporal aspect to the imagery, while the written references imply that habits and appearance are unchanging over time.[5]

Illus. 1.1. "An Exacter Figure of the Plantaine Fruit." Wood block after drawing by Thomas Johnson, *The Herball . . . Gathered by John Gerard*, ed. Thomas Johnson (London, 1633), 1516. (Courtesy of the Department of Special Collections, Van Pelt–Dietrich Library Center, University of Pennsylvania.)

Specific patterns in prose and imagery convey these special qualities of time and objectivity. In plant imagery, the object is "removed" from soil to show the root and is seen in side view to display the most information. In bird and mammal pictures, the specimen is also displayed in side view, while the invertebrates' dorsal (top) view is often displayed as the most characteristic aspect. Separated feet, beaks, eyes, or flowers may surface on the blank background to provide clearer details of characteristic parts. The title of the species is almost always printed on the page or closely associated with the image. A listing or catalog of species with accompanying descriptions (as in the travelers' accounts) is a major component of natural history discourse. The language within descriptions is usually "impersonal," in third-person narrative. In the imagery, the specimen depicted is assumed to represent the entire group; in prose, the entity most often is referred to as "it" or the composite "they." The physical characteristics frequently are listed in phrases with no verb ("eye brown, . . ."), further emphasizing the atemporal nature.[6]

The developments regarding classification further distinguish the discourse. Pioneers of systematic taxonomy such as John Ray in the sixteenth century began to base classification on the physical structure of eyes, feet, and beaks rather than on alphabetical order or more anthropocentric criteria such as utility to humans. Beginning in the 1600s, Latin was used to present separate terminology intelligible only to naturalists. Chunks of texts are devoted to erudite explanations of terminology and classification—in other words, why naturalists order certain species within lists. Carl Linnaeus in the mid-eighteenth century codified the practice of classification according to isolated physical characteristics by designating a species by two names, the genus and species, unlike previous strings of Latinate names. As soon as binomial nomenclature was accepted, naturalists sought a "natural" system based more on the organism's total structure well into the nineteenth century.

The genre was flexible enough to adapt to changes within the discipline and from the surrounding culture. The imagery and texts could respond to different taxonomic emphases by including varying visual details and more verbal description. As mentioned, early texts related commonly believed fables, while later ones included such human interactions as pet-keeping, gardening, hunting, and medicinal uses. Listings often were hierarchical, with humankind placed first in the "scale of nature" and invertebrates and plants last, in continuation of traditional Christian domination over the animal kingdom; yet the separate catalogs of sections of animal or plant kingdoms published in the eighteenth and nineteenth centuries implicitly gave these segments new weight and importance. Influenced by the humanitarian movement of the eighteenth century, more naturalists began to draw explicit parallels between humans and nonhumans. The inclusion of first-person observations particularly extended the discourse. The travelers' accounts, for example, related specific encounters with individual members of species and so interjected temporal and individualistic elements to the discussions. Collating these experiences widened the known range of actions and habitats for species and also recorded the multitude of human reactions to plants and animals.[7]

The Discourse of Travelers: Foreign and Native

Travelers used natural history discourse and the closely related conventions of the ethnographic "manners and customs" style to portray human and nonhuman inhabitants of new lands. Students of ethnography have especially analyzed how the European travelers and colonists portrayed native inhabitants as "the other," radically different beings with few similarities to their human experiences. Europeans described the costumes, homes, and appearance of many individuals as belonging to one entity, often a tribe. These characteristics were assumed to be unchanging, so that the inhabitants lacked the historical past and presumably the future that the Europeans possessed. The resemblances to natural history discourse are obvious. The travelers used this process of "othering"—the representation of other entities as fundamentally different yet with definable characteristics—to separate and thus order their perceptions of new and often confusing places and experiences. These conventions allowed them in their subsequently published accounts to communicate more easily the wealth of sensory perceptions to those with no knowledge of those entities.[8]

The counterbalance to the desire for separation from these "others" was the intense curiosity about the most unusual forms of natural life. Travelers through their oral and published tales became known for their credulous and marvel-seeking tendencies.[9] The flora and fauna riveting their attention were those that differed greatly from European species, such as the cacti, armadillos, porcupines, and opossums, for most North American species were remarkably similar to European analogues. Some preliterate cultures and some folktales in modern American popular culture may reflect uneasiness and even fear of such animals as the alligator and armadillo, which appear to be crosses between two types (the alligator between a snake and fish, the armadillo between a reptile and pig), but in the fifteenth to eighteenth centuries elite culture embraced the different. Desire to know "other" forms was lauded as a necessary component for human intelligence, as demonstrated in Edmund Burke's comment, "The first and the simplest emotion which we discover in the human mind, is curiosity." The authority given by firsthand accounts and the scientific guise of natural history discourse made travelers' descriptions of curiosities all the more appealing.[10]

Its numerous readers may associate one of the most famous American travel narratives, William Bartram's *Travels through North and South Carolina, Georgia, East and West Florida* (1791), most closely with the lyrical landscape passages of savannah groves and dramatic encounters with bellowing alligators that entranced the European Romantics.[11] Bartram also used natural history prose to capture and convey the wealth of new sensations he encountered in his trip through southern landscapes in 1773 to 1776. His heritage as the son of the colonial botanist John Bartram and his own career as a plant collector for European naturalists had given him the general tools of natural history, while the works of previous visitors in the region, particularly John Lawson's *A New Voyage to Carolina* (1709) and Mark Catesby's *The Natural History of Carolina* (1731–1743), provided him with the more specific knowledge of the identities of many species. For example, Bartram lists in chapter 10 of *Travels* the reptiles, mammals, and birds with their physical descriptions and habits, relating physical attributes in the standard impersonal style of natural history: "The largest frog known in Florida and on the sea coast of Carolina, is about eight or nine inches in length from the nose to the extremity of the toes." Most readers not unexpectedly remember his more fervid style.[12]

Pennsylvania Bartram could not have recounted the species residing in the southern climes without Lawson's and Catesby's extant texts. He indeed followed the common pattern of absorbing natural history knowledge through reading and viewing texts and correcting and expanding that knowledge through personal observation. John Lawson, a gentleman surveyor hired to explore the little-known Carolinas, covered roughly 550 miles in fifty-nine days in early 1701, and in 1709 published his travel account in England. His *Voyage* lists the fauna and flora with a completeness copied by his contemporaries and admired by later historians. His personal observations and interviews of Indian and European residents supplemented the previous bare-bones lists with richer information about habits.[13] Another Englishman, Mark Catesby, relied on Lawson about two decades later to provide a kind of checklist for listings and observations in his *Natural History of Carolina*. He, however, takes pains to present his firsthand observations as superior to those of Lawson when he differs. For example, Catesby includes a larger number of reptiles because, as he proudly emphasizes, he asked "several of the most intelligent persons" whether they knew of more serpents and "none of them pretended to have seen any other kind."[14]

Although Bartram was new to the southern climes, he, unlike his naturalist predecessors, John Lawson and Mark Catesby, had been raised in the colonies. His years of residing in America allowed him to correct his predecessors' observations, especially those of Catesby, who had covered the same territory in briefer trips from Great Britain. Bartram could clarify Catesby's mistake of the catbird having but one note ("he, in reality, being one of our most eminent songsters") because Bartram had heard catbirds "during their nuptials" in Pennsylvania while Catesby had heard them only in their winter residences in the South, "when they rarely sing." Bartram's residency proved crucial for his discussion of bird migration. He noted that because no one since the ancients knew where swallows lived in winter, it had been conjectured that the birds were hiding in caves or hibernating in the mud. That creatures so gloriously swift in air would choose not to migrate but to bury themselves deeply offended his belief in bird intelligence; therefore, Bartram was eager to publish his observation of huge flocks of swallows flying over the Carolinas in fall to destinations south and in spring to nesting places in Pennsylvania. He finishes the lengthy exegesis that he calls his "mite towards illustrating the subject of migration" with an exhaustive listing of the five categories of birds according to their migration and nesting patterns (yearlong residents in Carolina or Pennsylvania, winter farther south than Carolina but nest in Carolina, etc.), based on his observation in both regions. He explains why Europeans like Catesby had been silent on the subject: The knowledge "only could be acquired by travelling, and residing a whole year at least in the various climates from north to south."[15]

Bartram, like travelers before him, turned to natural history in his attempt to understand his ever-changing surroundings. Familiarity with the natural history of a species fostered his security in the new landscape, often in the obvious physical sense, as in the knowledge of venomous snakes, but also in the realm of psychic comfort and control of environs. Armed with his knowledge of the new Linnaean binomial nomenclature, he listed every plant species he remembered in his initial description of a place: "These hills are shaded with glorious Magnolia grandiflora, Morus rubra, Tilia, Quercus, Ulmus, Juglans, &c. with aromatic groves of fragrant Callicanthus Floridus, Rhododendron ferrugineum, Larus Indica, &c." Such listing mirrored his process of identifying familiar plants in an unknown territory as if he was seeking out markers. After a day of chasing away two bears and watching alligators, it was comforting to know the identity of the creatures stirring among the leaves as he drifted to sleep. Bartram gives a complete description of these rustlers:

> The wood-rat is a very curious animal. It is not half the size of the domestic rat; of a dark brown or black colour; its tail slender and shorter in proportion, and covered thinly with short hair. It is singular with respect to its ingenuity and great labour in the construction of its habitation, which is a conical pyramid about three or four feet high, constructed with dry branches, which it collects with great labour and perseverance, and piles up without any apparent order; yet they are so interwoven with one another, that it would take a bear or wild-cat some time to pull one of these castles to pieces, and allow the animals sufficient time to secure a retreat with their young.

Later Americans on their journeys through unknown territory would likewise rely on the knowledge and order afforded by natural history.[16]

The Development of Life Histories

In 1710 John Lawson had written to his English patron, James Petiver, who supported his scientific collecting efforts, outlining his intention to observe and correct more thoroughly the Carolina flora and fauna about which he had already written in *A New Voyage*. The goals he enumerated to Petiver demonstrated that Europeans and colonists well before the creation of the United States had desired the fullest information on the total life cycles, not only discovery and recounting of species. Lawson promised to record "all of plants I can meet withall . . . always keeping one of a sort by me giving an account of ye time & day they were gotten, when they first appearing, wt soil of ground, wn. the flower seed & disappear & wt. individuall uses the Indians or English make thereof & to have it enough of the same." For birds, "to know if possible the age they arrive to, how & where they build their nests, of what material & form, the coulour of their eggs and time of their Incubation & flight, their food, beauty & colour, of wt. medicianall uses if any. if rarily designedd to the Life, this would Illustrate such a history very much, their musical notes & cryes must not be omitted, wch of them abide with us all ye year." In the case of insects, the points of concern were "the months they appear to us in the places of their resort, how they breed & wt. changes they undergo."[17] Such detailed accounts would transcend the existing brief listings and static physical descriptions. Lawson was killed in an Indian attack before he fulfilled these ambitions; it would take many individuals living year round in the colonies decades to collect and collate such observations. Ironically, the land no longer would belong to Europe by the time of the most sustained observation.

John Abbot

One such researcher was John Abbot (1751–1840?) who lived in the lowlands of Georgia. Abbot, like Catesby and Lawson, came to America as a natural history collector working for other British naturalists in 1773 and eventually settled near the Savannah River. Unlike Lawson and Catesby, he remained for some sixty years. As a youth, he had modeled his watercolor style on the plates depicting insects and birds in the works of two premier English naturalist-artists, Eleazer Albin and George Edwards, and had met and received praise from Edwards. He subsequently made over two thousand drawings in America and often sold sets to European collectors in lieu of specimens, with his personal observations in the margins. For his bird drawings, closely modeled after those of Edwards, he recorded the songs and habitats in such phrases as (for the yellow-billed cuckoo): "Rain Crow—From their Cry being said to be a sign of Rain, Frequents Oak Woods and Swamps. Is a Bird of Passage appearing in the Spring."[18]

Abbot's most original contribution relates to entomology. Over the decades, he studied particularly butterflies and moths in their outdoor habitats and inside his home. From his field studies, he learned the favorite food plants of the individual species and took home eggs and pupae to be hatched under his own eyes. He compiled the times and manners of the various species' metamorphoses, as for the Luna moth: "Feeds upon Sweet Gum, Walnut, Hickory, Persimmon, &c. Spun up 31st May. Brd. 18th June. Spun up 23rd June, brd 10th July, Spun up Spt. 6th. . . . Continues breeding all the summer. [*Spun up* signifies that the caterpillar is spinning its cocoon.]" European collectors had previously reared butterflies from eggs or cocoons in order to procure fresh specimens, but in contrast to Abbot, they had neither as thoroughly noted transitional states nor observed their breeding in the wild.[19]

Abbot's visual iconography expressed his knowledge of the species' characteristic life cycle (Illustration 1.2). He arranged the branch or blossom of a favorite food plant, the adult (male and female if visually different), the caterpillar, and the pupa (in the cocoon) on this same plant to represent the stages. European authors such as the great seventeenth-century illustrator Maria Sybilla Merian in her *Insects of Surinam* (1705) and Eleazar Albin had shown food plants and cocoons earlier but not so frequently or with so many life stages as Abbot. His years of study and access to specimens enabled this consistency. English naturalist James Edward Smith edited Abbot's manuscript and published two hundred drawings for *The Natural History of the Rarer Lepidopterous Insects of Georgia* (1797), while an English engraver, John Harris, engraved the plates and probably arranged for the hand coloring in London. Smith carefully left Abbot's notations on food plants and metamorphoses in the text, but he had to add the scientific nomenclature himself, as Abbot claimed he lacked this skill.[20] The French naturalist Jean Alphonse Boisduval and the American entomologist John E. LeConte published more Abbot butterfly drawings, which the artist had sold to LeConte, in the *Histoire générale et iconographie des lépidoptères et chenilles de l'Amerique* (Paris, 1829–1837).

Alexander Wilson: Ambition and Fulfillment

Alexander Wilson created the first fully illustrated original natural history monograph published in the United States, *American Ornithology* (1808–1814). (Frenchman Louis J. P. Vieillot's notes on birds he observed when he lived in America were published in Paris as *Histoire Naturelle des Oiseaux de l'Amérique Septentrionale* in 1807, and William Barton's 1803 *Elements of Botany*, for example, held only a few engravings after William Bartram's drawings.) Wilson in the prospectus issued in 1807 gave full play to his groundbreaking achievement: "Nothing similar to the present undertaking has ever been attempted in America." What is most impressive about Wilson's ambition is the extent to which he fulfilled it. After the author provided a list of some two hundred men of first-rate credit who pledged to buy the work, Philadelphia publisher Samuel Bradford was persuaded to undertake the project. Alexander Lawson (and to a lesser extent, other craftsmen) engraved Wilson's hundreds of original drawings between 1807 and 1813. When he was not traveling to gain subscribers, Wilson supervised an ever-changing crew of relatives and young adults who colored the thousands of plates; he colored them himself when necessary. His extensive descriptions of 264 species, in many cases several folio pages in length, exceeded those of his European predecessors while he added over forty new species. The years of intense study, drawing, writing, and traveling probably contributed to Wilson's death in 1813 after eight volumes had been written. (George Ord, a Philadelphia member of the Academy of Natural Sciences, completed the ninth volume from Wilson's remaining drawings and notes.)[21]

Illus. 1.2. "Sphinx Carolina. Tobacco Hawk-Moth." Hand-colored engraving by John Harris after John Abbot, for James Edward Smith, *The Natural History of the Rarer Lepidopterous Insects of Georgia* (London, 1797), vol. 1, Plate 33. (Courtesy of the Winterthur Library, Printed Book and Periodical Collection.)

Wilson's early career made him an unlikely candidate for orni-

thological fame. Unlike Catesby and Abbot, the largely self-educated
son of a Paisley weaver had shown no early love for natural history
but instead wrote poetry in the vein of Robert Burns, which he sold
while peddling. After two brushes with the law regarding libel charges
stemming from his satirical poetry, he emigrated from Scotland with
a nephew to America in 1794. He eventually settled in the Philadelphia area, whose nascent literary culture nurtured his ambitions.
Samuel Bradford offered him the editorship of the American edition
of *Rees Cyclopedia,* and local papers and a leading magazine, *The Portfolio,* published his new poems steadily. These poems, including the
epic (at least in its length of over two thousand lines) *The Foresters,*
conveyed a not uncommon love of romantic scenery, expressed
through his persona of the lonely wanderer in "sweet rural scenes . . .
where Nature's charms in rich profusion bound."[22]

A newfound interest in ornithology, stimulated by the mentor-
ship of William Bartram and solidified by the study of natural history
books, transformed the conventional poet into a naturalist and prose
writer of originality. In 1803 he met Bartram while living near the
elder man's estate, "Bartram's Gardens" in Philadelphia, and the naturalist lent him his copies of the European authors. A study of ornithological literature including Bartram's own works, George
Edwards's books, and the English edition of Linnaeus's *System Naturae*
shifted Wilson's ambition to writing natural history. Wilson learned
natural history principles and content by constant rereading. Ham-
pered by lack of education in the classical languages, he struggled to
understand the Latinate binomial nomenclature. Despite these hur-
dles, he so thoroughly learned the previous authors' descriptions that
he could refer to and often correct comments on identity and distribution in *American Ornithology.* Historians of reading have discussed
the role of such "intensive reading" in various cultural contexts from
sixteenth-century Bible reading to nineteenth-century consumption
of novels.[23] Wilson's adult autodidactism exemplifies the transmission
of one form of literary culture through an individual's efforts.

Wilson likewise learned drawing by studying finished products
rather than through formal education. In 1803, the same year in
which he met Bartram, he began to learn natural history illustration
by copying prints owned by his mentor; once, after copying a botanical plate, he moaned in mock-tragic horror, "Sir, I have murdered
your Rose." Bartram and his niece Nancy, whose own ornithological
illustrations adorned her uncle's articles, continued to encourage
Wilson's efforts to draw from actual specimens in the standardized

poses of natural history. Wilson, a schoolteacher by day before his editorship, drew at night by candlelight. The one advantage of teaching was the willingness of his students to bring specimens: "I have had live crows, hawks, and owls—opossums, squirrels, snakes, lizards, etc., so that my room has sometimes reminded me of Noah's ark."[24]

From the Renaissance encyclopedias onward, it was standard for discussions of habits to follow the naming and anatomical description of the species. Wilson so amplified these accounts that they became the full "history" of the species. Like Bartram before him (and John James Audubon after him), Wilson juxtaposed dry physical details, emotional dramas, interesting prose pictures, and, in a few cases, even his own poetry. The length (sometimes several pages per entry) indicates the totality of information, as Wilson included much of his promised "places of resort, General Habits, Peculiarities, Food, Mode of Constructing their Nests, Term of Incubation, Migration, &c. &c." The descriptions open with the scientific names given by past naturalists, a practice known as "synonymy." They amply describe characteristics of each species deemed significant, which, given Wilson's fascination, meant seemingly every detail he knew. He included impersonal listings of physical descriptions of markings including eyes, beaks, wings, and feet, not only for the adult male but also for the female and young; he also provided information attributed to other authors.

Through concentrated study both in- and out-of-doors, at home and beyond, Wilson accumulated the substance of life histories to an extent hitherto unknown. For example, he studied over fifty orchards frequently by red-headed woodpeckers to judge whether they injured trees by sucking their sap as commonly believed; contrary to popular belief, those trees with the most woodpecker holes were the healthiest, an observation he linked to the woodpeckers' appetite for the insects that otherwise would infest the trees. In order to ascertain the proportion of grain to insect larvae, he dissected over fifty bluebirds. Wilson found as fledglings and raised from eggs birds of many species to observe their habits, foods, song, and plumage changes over time and at intimate range. Keeping a box next to his window for a pair of nesting wrens gave him unique insight into mating habits. After a cat killed the first female, Wilson observed that the male appeared confused, but a few days later a second female came as his mate to nest. He was ever alert to bird life on his frequent field excursions around Pennsylvania and broader travels (he once boasted that he tramped through muddy fields for four hours in an attempt to see a nuthatch). While walking home from a strenuous journey to Niagara, singing chanteys to drown out his younger companions' complaints about the rain and rough roads, he managed to espy and shoot "a nondescript jay." In Louisville, he even bird-watched with his future successor, John James Audubon; and at the William Dunbar estate outside Natchez, Mississippi, he observed and shot his first Mississippi kite, his later description of its flight memorable: "In my perambulations I frequently remarked this Hawk sailing about in easy circles, and at a considerable height in the air, generally in company with the Turkey Buzzard, whose manner of flight it so exactly imitated as to seem the same species, only in miniature or seen at a more immense height."[25]

Many similar passages in *American Ornithology* display Wilson's command of language in service of bird portrayal. He often ascribes human motivation to birds, and in the bluebird description offers an image contributed by William Bartram of the male bluebird "courting the female": "It is pleasing to behold his courtship, his solicitude to please and to secure the favor of his beloved female. He uses the tenderest expressions, sits close by her, caresses and sings to her his most endearing warblings. When seated together, if he espies an insect delicious to her taste, he takes it up, flies with it to her, spreads his wings over her, and puts it in her mouth." Wilson adds his own portrayal of jealousy: "If a rival makes his appearance,—for they are ardent in their loves,—he quits her in a moment, attacks and pursues the intruder as he shifts from place to place, in tones that bespeak the jealousy of his affection."[26]

In many other passages, Wilson employs a wide range of adjectives and nouns often associated with humankind to picture vividly distinct avian actions such as flight, song, and movement on the ground. The heron's step is "stately," and the Mississippi kite "sails in easy circles." He saw a cardinal raise a cow bunting fledgling, so his familiarity infuses the description of it uttering its unlovely song with crisp visual and aural detail:

> He [the cow bunting] spread his wings, swells his body into a globular form, bristling every feather in the manner of a Turkey cock, and, with great seeming difficulty, utters a few low, spluttering notes, as if proceeding from his belly; always, on these occasions, strutting in front of the spectator with great consequential affectation. To see the Red-Bird [cardinal], who is himself so excellent a performer, silently listening to all this guttural splutter, reminds me of the great Handel contemplating a wretched catgut scraper.[27]

Study only increased his fascination with birds. As he expresses in one *American Ornithology* description, "those who have paid the most minute attention to their manners, are uniformly their advocates and admirers." Birds' dual association with humanity and their uniqueness or "otherness" delight him. He praises the devotion shown by bird parents as worthy of human emulation, and his actions with the birds he raised blurs the traditional distinctions between wild and tame, outdoors and indoors, and human and nonhuman. He fed them by hand or from his mouth; they sometimes ranged freely in his rooms and picked up crumbs from his mouth and clothes. After Wilson rescued a young hummingbird stunned by its fall from a nest, "a woman in his house" pressed it to her bosom to revive it and "dissolved a little sugar in her mouth, into which she thrust its bill, and it sucked with great avidity." The very exoticism of unusual colors and tropical origins stir him, as he expressed in the prospectus: "On this hand he beholds one species just returned from the orange groves of Guiana and Surinam. . . . while a brilliant and numerous family, in green, blue and glowing scarlet, that but lately caught the eye of the sun-burnt savage of Brazil, amidst his native woods, now flutters through our orchards." Knowledge of their tropical migration patterns enhanced, not demystified, their specialness.[28]

The doctrine of natural theology—the belief that the natural world manifested God's intentions—had been circulated through Anglo-American literate culture decades before Wilson, as eloquent proponents such as Bartram throughout the *Travels* praised the study of natural history as revealing God's handiwork. Wilson's personal study of birds' intricate structures and remarkable habits convinced him afresh of God's perfect workmanship. For example, he elucidated the habit of the cow bunting (or cowbird) female, who laid her egg in the nest of other species instead of building her own. The bunting egg hatched a day or two earlier than the other species' eggs, thereby ensuring its survival. The other eggs "disappear," crowded out of the nest: "In this singular circumstance, we see a striking provision of the Deity; for did this egg require a day or two more, instead of so much less, than those among which it has been dropped, the young it contained would in every instance most inevitably perish; and thus, in a few years, the whole species must become extinct." The barred owl's feathers may seem an unlikely topic for a paean to God's creative abilities, but Wilson's discernment of how the various different types adapt the owl to its job of stealthy hunting moved him to praise:

Those [feathers] that surround the bill differ little from bris-
tles; those that surround the region of the eyes are exceedingly open, and unwebbed; these are bounded by another
set, generally proceeding from the external edge of the ear,
of a most peculiar small, narrow, velvety kind, whose fibres
are so exquisitely fine, as to be invisible to the naked eye . . .
the webs of the wing-quills are also of a delicate softness, covered with an almost imperceptible hair, and edged with a
loose silky down, so that the owner passes through the air
without interrupting the most profound silence. Who cannot
perceive the hand of God in all these things?[29]

Wilson's bird study revealed other habits among birds, not always
admirable or expected. The natural history discipline required every
aspect to be recorded, and not all birds received praise as devoted
mates or parents (the mating of the cow bunting was "concubinage"
because they did not pair off to nest). Observation could reveal differences from conventional human behavior, as in the female eagle
being larger and "more brave" than the male, "a circumstance common to almost all birds of prey." The eagle mother was devoted in
bringing food to her young, but the decaying fish made the nest pu-
trid. There could be intraspecies war, as in the titmice "known to
attack young and sickly birds," Wilson admitted, after he killed a specimen whose head showed a dent in its cranium, probably from another titmouse bashing it. In another instance, a caged male cardinal
killed the female, an action Wilson attributed to jealousy over her
superior singing.[30]

The Advantages of Residency

Wilson expressed his love for the country that encouraged his
aspirations in outright boasts and flagrant taunts throughout *American Ornithology*. More important for American cultural achievement,
his exploitation of the advantages given to an inhabitant naturalist
changed the natural history format's text and imagery and established a precedent for illustrated monographs in the United States.
His personal observations, coupled with those of other Americans
whom he recruited for the mission, so greatly enhanced the life history portion of the traditional species description as to transform it.
Likewise, his experience seeing the subjects in their environment
over an extended period inspired him to emphasize their distinctive
movements through artwork.

Desiring that the work produced in America succeed on European terms, he boasted that the homegrown "elegance" of typesetting, binding, and paper quality was "not inferior" to English rivals. In this era of fervid promotion of U.S. manufactures, Wilson went so far as to apologize for the imported watercolors but took pains to note that the rags from which the paper was made were American. The plates were etched and hand-colored just as in great seventeenth- and eighteenth-century natural history folios. The format permitted him life-size images (like some illustrations by Mark Catesby and George Edwards, another British naturalist) in many cases, but Wilson included too many birds to give most a separate plate. Alexander Lawson, a Scottish émigré, undertook the task of translating Wilson's original watercolors into engravings. The extreme precision required in copying the exact measurements and mimicking the delicate pencil lines of the original drawings was as great a challenge as faced any engraver in the early republic.

On the theoretical battleground, Wilson bravely undertook the task of assigning scientific nomenclature following the outlines of British ornithologist John Latham, although he had no prior knowledge of classification or Latin. Wilson once confessed to Bartram that the scientific names in the latest edition of Linnaeus's work made little sense to him because they did not consistently signify the distinctive character of the species; for example, the term *migratorius* could apply to many birds equally well as to the robin. Each synonymy begins not only with the Latinate name assigned by other authors but also with his new Latinate nomenclature if he considered the previous nomenclature to indicate incorrectly the bird's relationship to nearby species. Wilson asserted, despite his recent initiation into natural history, that in some respects his field expertise rendered his classifications superior to European "closet" naturalists "working from mummifed skins." Observation clarified previous European errors of identifying female or young of one species as a separate species, mistakes arising from working only from dead specimens, and the knowledge of habits led to better groupings of similar birds than reliance on "mere examination of physiognomy."[31]

The ample life history descriptions following his synonymy sections were the element in which Wilson truly improved upon his predecessors. The quantifiable aspect of length and the qualitative aspects of content and quality are generally superior to earlier descriptions. His years of residency and greater access to specimens, for example, allowed him to postulate the economic value of insect-eating birds to farmers (from examinations of many bird stomachs), thus foreshadowing economic ornithology. He also had the advantage of a considerable corpus of knowledge about American birds through Catesby, Edwards, Turton (the English edition of Linnaeus), Latham, and Bartram to use as a scaffold for his contributions. He did not have to originate physical descriptions and names for the majority of birds, for example, and, like Catesby and Bartram before him, he could build on others' descriptions. His ability to illustrate in writing both his individual impressions and the general aspects of bird life imparted a readability and freshness to the text.

The Europeans' literature gave an American prime opportunity to criticize them as well. Wilson's delight is evident in his mockery of the eminent British naturalist John Latham on the bluebird: "But whoever informed Dr. Latham, that 'this bird is never seen on trees, though it makes its nests in the holes of them!' might as well have said, that the Americans are never seen in the streets, though they build their houses by the sides of them." Wilson, no doubt influenced by the eighteenth-century Scottish school of "common-sense" thought, thus frequently invoked everyday good sense and the reliance on firsthand observation. What most offended Wilson was the European imputation (often attributed to the Comte Georges-Louis Leclerc de Buffon, the eminent author–natural historian of the monumental *Histoire Naturelle*) that American birds had degenerated over time from European prototypes. The proposition outraged both his patriotism and his admiration for the creatures he studied. The Creator had not mistakenly made any creature "rude, defective, or deformed." The noted American songsters such as the cardinal "were not inferior" to the nightingale, claimed Wilson, who had "listened a thousand times to both." In response to the degeneracy theory, Wilson reasoned that God had created similar species on different continents just as he had similar species living in the same locale: "[Is it], therefore surprising, that two different species . . . should have certain near resemblance to one another without being bastards?"[32]

Wilson and His Correspondents

Wilson fully exploited his advantage of bird informants. European naturalists such as Linnaeus and the English botanists had relied on American correspondents for specimens and information, but Wilson built even more such relationships through his proximity. In his travels, he broadcast his desire for specimens and information about habits to every acquaintance and advertised his project in newspapers, boasting "that not scarcely a wren or tit shall be able to pass along New York to Canada, but I shall get intelligence of it." Wilson's enthusiasm and persistence persuaded individuals not previously enamored of bird study to contribute information and explore the subject themselves. For example, residents of Easton, Pennsylvania, told him that swallows congregated in the town hall's chimney. Wilson was eager to refute the suspicion that swallows hibernated in trees and chimneys instead of migrating, so, obligingly, Abraham Hart, Mordecai Churchman, and a "Mr. Arnott" actually looked in the chimney at Wilson's request in spring and winter to ascertain whether the swallows lived there. Churchman, a Quaker banker, went so far as to ask fellow Eastonians whether they had noted nests or swallows wintering in the chimneys, and communicated his results to Wilson. (Apologizing to Churchman for the imposition, Wilson joked, "Truth they say sometimes lies at the bottom of a well—It may also lurk at the bottom of a Chimney.")[33]

Some of Wilson's correspondents were talented naturalists (although not as ambitious in publications as Wilson) and in several instances sent life histories worthy of Wilson himself. Samuel Latham Mitchell, senator, proto-ethnologist, and author of *The Fishes of New York,* contributed almost the entire text for the pinneated grouse, including the interesting category "Amours," featuring its drumming and blowing bladderlike appendages to attract the female. Nathaniel Potter, a prominent Baltimore physician who had been a student of Benjamin Rush at the University of Pennsylvania's medical school in the 1790s, in 1809 displayed his longstanding passion for and knowledge about birds by responding to Wilson's plea in the *Portfolio* for information about a specific species. He, like Wilson, had heard that the cow bunting did not nest. Years earlier, while residing on Maryland's eastern shore (well supplied with that species), he had offered a premium to anyone who could find such a nest, and "the negroes in the neighborhood brought me a variety of nests; but they were always traced to another bird." His quest after a cowbird ready to lay eggs embodies the naturalist's determination:

> If a discovery suitable to her purpose cannot be made from her stand, she becomes more restless, and is seen flitting from tree to tree, till a place of deposit can be found. I once had an opportunity of witnessing a scene of this sort, which I cannot forbear to relate. Seeing a female prying into a bunch of bushes in search of a nest, I determined to see the result, if practicable; and, knowing how easily they are disconcerted by the near approach of man, I dismounted my horse, and

proceeded slowly, sometimes seeing and sometimes losing
sight of her, till I had travelled nearly two miles along the
margin of a creek. She entered every thick place, prying with
the strictest scrutiny into places where the small birds usually
build, and at last darted suddenly into a thick copse of alders
and briers, where she remained five or six minutes, when she
returned, soaring above the underwood, and returned to the
company she had left feeding in the field. Upon entering the
covert, I found the nest of a Yellow-Throat, with an egg of
each.

(Presumably he was able to find his horse.) Potter could not check
on incubation time as he wished, because both eggs were missing on
his next visit. He speculates on why the bunting eggs hatch earlier
than the original eggs, but in a quintessential statement of observational empiricism claims that "the facts remain, whatever the theories." Potter mentions that during his stay on the eastern shore he
kept copious notes on birds, but upon removal to that worldly city,
Baltimore, "artificial ideas have occupied the place of his impressions
of nature." His medical research style indicates a similar intrepidity
and hands-on approach, for he inoculated himself with the perspira-
tion of a yellow-fever patient to prove that the disease was not conta-
gious.[34]

The Iconography of the Living Bird
Wilson's artistic style as represented in the plates drew heavily on
his inhabitant advantages. Because of his correspondents and own
abilities in capturing birds, he could usually choose a "fine" dead
specimen or even a live one from which to draw and could include
the young or female of the species if they had been confused. This
self-taught artist modeled his drawing strictly after an individual example instead of generalizing from a group of specimens, seeking the
quintessential example in coloring and health to represent the species. (The eagle who served as the *American Ornithology* model, for
example, was a "beautiful female.") His drawing technique and Lawson's subsequent engraving placed a premium on great specificity of
certain parts such as the eyes, legs, nostrils, because these details were
thought to differentiate species. Wilson's knowledge and careful supervision of plate coloring guarded against mistakes like that of Englishman George Edwards, who had painted plates of the chestnut-
sided warbler "blood red."[35]
The text emphasized the times in which he drew from the living

bird. His drawing of the Mississippi kite was "made with great care from the living bird," a bird so full of life that it pierced Wilson's wrist with his hind claw as "its dark red eye sparkled with rage." He was proud of his intention to capture, in his predecessor Mark Catesby's words, the "gestures peculiar to every kind of bird," as in the kite's stance and the cedar waxwing's erect crest. Wilson, if supplied with the living bird, most likely made a sketch of its attitude, then perhaps killed it to measure and render more precisely the minute details of feet and nostrils in the finished drawing. For example, when he saw and heard the courting male pinneated grouse in the Kentucky barrens "tooting" by blowing out the bladderlike material around the throat, "so very novel and characteristic did the action of these birds appear to me at first sight, that instead of shooting them down, I sketched their attitude hastily on the spot, while concealed among a brush-heap, with seven or eight of them within a short distance." He assures the reader/viewer that the specimen depicted in the plate "is most perfect of several elegant specimens shot" (Illustration 1.3). Wilson (and his engraver Alexander Lawson in the volume published posthumously) placed the bird against a typical background drawn from their visual memories, thus portraying "the living animal in nature."[36]

Even when Wilson worked from a dead specimen, his knowledge of the species' typical movements informed his interpretations. Europeans like Mark Catesby and George Edwards had attempted movement and background (Illustration 1.4), and the noisy blue jay from Catesby's *History* is one successful example (Illustration 1.5). They could not convey as consistently as Wilson actions like the meadowlark's crouching near the ground or the nuthatch's securely running up a tree upside down. "Field experience" became integrated into ornithological illustration.[37]

The Audience in the Early Republic

Wilson called his ambition of publishing a work in the European folio tradition "quixotic," conceding, "I have been so long accustomed to the building of airy castles and brain windmills." His trepidation is understandable. Expensive illustrated books like Wilson's needed commitment (including steady cash infusions) from subscribers to fund the progress of the work, just as in Mark Catesby's day (and in the case of Audubon's huge folios to come), because publishers did not possess the capital necessary to pay for the labor and materials. Each volume was priced at $12, with the projected final price for the set $120. The sum far exceeded the price of contemporary novels, priced between fifty cents and one dollar, and was even more expensive than most portrait paintings; only the multivolume, heavily illustrated encyclopedias also published by the subscription method were comparable in ambition and cost in the early republic. His potential audience's willingness to pay for natural history books was untested, unlike the well-established patronage base in Europe.[38]

Illus. 1.3. "Pinneated Grouse, Blue Green Warbler, Nashville Warbler." Hand-colored engraving by Alexander Lawson after Alexander Wilson, for Wilson, *American Ornithology,* vol. 3, Plate 27. (Courtesy of the John Work Garrett Library of the Johns Hopkins University.)

Collecting the Subscribers

Wilson's American contemporaries, moreover, compared with their foreign counterparts, had spotty knowledge of illustrated works of natural history, let alone those concentrating upon ornithology. Botanical illustration was best known through the inexpensively priced herbals transplanted from the Germanic tradition and through eighteenth-century English works such as Philip Miller's *The Gardener's Dictionary* and the *Curtis Botanical Magazine,* found in wealthy private libraries like that of George Washington. Zoological works by Catesby and Edwards and the Comte de Buffon's tome, *Historie Naturelle,* were accessible only in a few private libraries and public

Illus. 1.4. "The American Water Rail." Hand-colored engraving by George Edwards, for Edwards, *Gleanings of Natural History* (London, 1754), vol. 2, Plate 279. (Courtesy of the Library, Academy of Natural Sciences.)

collections such as that of Harvard University. Natural history information was available, however, through more-popularly priced (albeit still expensive) forms. British Oliver Goldsmith's adaptation of Buffon first published in 1774, the *History of Animated Nature,* was imported in many versions and even published in the United States, all editions light on scientific nomenclature but heavy on interesting habits. The wood engravings in Englishman Thomas Bewick's 1790 *History of Quadrupeds,* originally intended for children, so inspired the young New York doctor Alexander Anderson, who first viewed them in 1794, that he learned wood engraving in order to copy the work. The few American illustrated magazines followed their English predecessors' lead in including natural history snippets as part of a well-rounded selection of the arts and sciences. The articles and engravings, although on American subjects like the moose and the buffalo, were adapted from English sources with more or less accuracy depending on the quality of original design and skill of the American engraver (Illustration 1.6).[39]

Illus. 1.5. "The Crested Jay" [Blue Jay]. Hand-colored engraving by Mark Catesby, for Catesby, *Natural History of Carolina, Florida, and the Bahama Islands,* vol. 1, Plate 15. (Courtesy of the Winterthur Library, Printed Book and Periodical Collection.)

Illus. 1.6. "The Buffallo." Engraving by anonymous artist, for *Massachusetts Magazine* 4 (April 1792), frontispiece for April number. (Courtesy of the Winterthur Library, Printed Book and Periodical Collection.)

Therefore Wilson knew his subscribers-in-waiting probably had been exposed to natural history and perhaps had enjoyed the taste. Even persons without a sophisticated background in European aesthetics could marvel at the realism and detail in Wilson's plates as he "shopped" them: Wilson bitterly told of a man who haughtily refused to buy because he "saw no use in them," but "this same reptile" could not deny "the beauty." Those who did know European plates would doubly appreciate Wilson's effort. Because Wilson knew that the bulk of his patrons would not be naturalists interested in the anatomical differences between species' feet, he appealed in his prospectus and introduction to each volume to a generalized feeling of public duty and civic pride, an unsurprising strategy given his patriotism and the extremely self-conscious nationalism in the first two decades of that century. He proclaimed not only the innate fitness of Americans describing native natural productions but the *Ornithology* as an American product itself. His prospectus and advertisements stressed natural history's ability to provide "elegant and rational amusement . . . beyond the idle prattle of novels," in tune with the concern for morals in the purer, better young republic. In his private correspondence, he identified his target audience as the acknowledged economic and cultural elite of their areas, persons of "wealth and taste." Wilson transformed the social system of receiving letters of introduction into a method of obtaining subscriptions, as supporters gave him letters to likely buyers. When Wilson lacked such letters

in Charleston, the librarian kindly gave him a list of the gentlemen
upon whom he should call.[40]

Wilson's final list of 450 subscribers contains names still noted
for their philanthropy and support of arts and sciences. Daniel Wadsworth of Hartford, Connecticut, who himself wrote articles on geology and later bequeathed his art collection to the public, subscribed,
as did President Thomas Jefferson, with whom Wilson had shared his
earliest bird drawings. Another subscriber was Philadelphian Nicholas Biddle, the one-time editor of the *Portfolio* before his tenure as
Bank of the United States president. John Collins Warren Jr. and Sr.
were prominent Boston doctors; the son later headed the Boston Society of Natural History. William Dunbar of Natchez offered gracious
hospitality although already in his final illness. Subscriber DeWitt
Clinton also wrote favorable reviews. The geographical distribution
of the patrons from New England to Kentucky, Georgia, and New
Orleans reflects not only the extent of Wilson's subscription trips but
also the pervasive sentiment of national pride. The varied sources of
income—from drug firms and merchant houses to the China trade—
hint that there was ample wealth to promote culture, given the right
object and a persuasive salesman like Wilson performing "direct marketing."[41]

The years of *American Ornithology*'s creation marked some of the
bitterest factionalism in American political history, as the major parties disputed over the 1807 Trade Embargo Act. A telling factor in
the subscription list is the presence of the most prominent Federalists
and Republicans. Wilson, an ardent Jeffersonian, naturally solicited
others of his political ilk, but *American Ornithology*'s appeal to the leading members of the other party, such as Clinton and Rufus King,
marks the significance of natural history in the national culture. Wilson himself tactfully defined its role of deflecting tensions and tran-
scending political allegiance in the introduction to volume 3:
"[Natural history books] are generally welcomed by people of all parties. They may be compared to those benevolent and amiable individuals, who, amidst the tumult and mutual irritations of discordant
friends, kindly step in to reconcile them to each other, leading the
discourse to subjects of less moment, but of innocent and interesting
curiosity till the mind forgets its perturbations and gradually gains its
native repose and composure. So comes, in these times of general
embarrassment, dispute, and perplexity, the peaceful, unassuming
pages of American Ornithology." The project apparently satisfied the
requirements of the common culture hoped for in the new republic,
shared by both political elites, and articulated in countless newspa-

pers and orations: godliness, public-spiritedness, nationalism, and so-
briety. Wilson successfully reached out to available patrons of the
opposing factions who themselves wished to claim the honor of patronage, notwithstanding politics.[42]

Naturalists and Non-Naturalists

Wilson had a skill of drawing out observations from persons who
were not bird lovers, such as the Eastonians who became willing chimney swallow watchers. Despite his debts to such helpers, he saw a distinction between the comparatively few "careful and intelligent"
naturalists like John Abbot (who met and supplied Wilson with specimens) and Nathaniel Potter, and the rest of humanity; in a few passages in *American Ornithology* and in his personal letters he firmly
articulated the division. The naturalist studied all species by the same
method while the non-naturalist adopted different relationships with
various forms. Sportsmen, for instance, knew very well only the species they hunted, for obviously utilitarian reasons. Wilson confided in
a letter to subscriber William Dunbar that, "so minute and exact is
Nature in her nice discriminations of different species of Birds and
so general vague and incorrect are almost all the accounts of illiterate
sportsmen . . . that I pay little attention to any accounts received from
such people relative to my pursuits. Nine times in ten they will be
found altogether erroneous." Few members of the populace recognized, let alone observed on a regular basis, the woodland or swamp
birds not hunted for food. True, common and distinctive species such
as the wild turkey, bluebird, and Baltimore oriole were deemed favorites, but only the naturalist would know (or care) about the "no less
than five species of these birds [wrens] in Pennsylvania, that, by a
superficial observer, would be taken for one and the same, but between each of which nature had drawn strong, discriminating, and
indelible lines of separation." It was the naturalist who "pointed out"
these lines through scientific species descriptions. Future naturalists
were to restate Wilson's comments on these differing relationships to
the natural world.[43]

The difference between common designations and scientific nomenclature embodied the two frames of thought. Wilson so de-
spaired of the variety of common or "local" names for a species
throughout the country that he asked correspondents to send him
dead specimens along with observations to enable him to discern the
species under discussion. Settlers in various locales created names
related to similarity to other species, color, and special habits or food;
thus, the Baltimore oriole was also known as the golden bird or firey

hang bird, and the red-winged blackbird was the swamp blackbird, marsh blackbird, or corn or maize thief. Scientific nomenclature, in contrast, strove to universalize names to facilitate international communication (although because of ever-changing taxonomic systems, Latinate names, ironically, were in constant transition).[44]

Wilson especially sought to dispel "foolish superstitions" that the non-naturalists of all ethnicities and classes attached to some species. Indians, for example, related different stories of the whip-poor-will, which they associated with the night: "Night, to minds of this complexion, brings with it its kindred horrors, its apparitions, strange sounds, and awful sights; and this solitary and inoffensive bird, being a frequent wanderer in these hours of ghosts and hobgoblins, is considered by the Indians as being, by habit and repute, little better than one of them." Wilson lumps the Indians' ignorance with that of "an illiterate German, or Scots Highlander, or the less informed of any other nation," as he scorns the old English custom of hanging up a dead kingfisher to ward off evil spirits. Literate persons also perpetuated "myths" that Wilson excoriates. Farmers' widespread killing of certain birds for their supposed fruit- and grain-picking could be counterproductive, warns Wilson, who studied the importance of insect-eating by dissecting birds' stomachs and noting their contents. The disciple of Bartram mocked the belief in hibernation of swallows (hence his chimney investigations), for the swift flight of birds alone made it improbable that they would choose to hide in the mud or lurk for months in torpidity: "Can a bird, whose vital functions are destroyed by a short privation of pure air and its usual food, sustain for six months, a situation where the most robust man would perish in a few hours or minutes? Away with such absurdities!" Robust "common sense" or the willingness to reason or listen to fact could erase these superstitions, because "all these people, are not so credulous: I have conversed with Indians who treated these silly notions with contempt."[45]

Birds carried to literate Americans centuries of European associations derived from medieval bestiaries and the subsequent Renaissance emblem books. For example, the pelican, reported to feed its young from its breast, became a symbol of parental affection in the medieval bestiaries which equated physical objects with heavenly messages. New World creatures became part of the emblem books, in which a moral phrase or story can be deciphered from a poem or message coupled with an image. Such books portraying by picture and story one aspect of animal life significant to human behavior were available in the early republic. For example, *Choice Emblems serv-*

ing to display the Beauties and Morals of the Ancient Fabulists, published in Philadelphia in 1793, presents a woodcut of a stork carrying its parent as "emblem of filial duty and affection" because "it is reported that their young ones assist the aged storks thus." Turtledoves are symbols of constancy, as they supposedly pine away when their mates die; the woodpecker epitomizes industry because of the pains it takes to whittle insects from bark.[46]

William Bartram also did not feel the modern hesitancy to attribute human gestures and attitudes to animals. He voiced his appreciation of the "ingenuity" of the "castles" of the wood rats and apparent reasoning powers of other birds and animals; similarly, he praised the coachsnake who accompanied him on a road, "now and then looking at me in the face." In other letters and articles, he openly calls upon humanity to attribute reason and moral sensibility to animals on account of such empirical observations. The earlier Cartesians' portrayal of animals as soulless automatons ruled only by instinct deeply offended him and contemporary English humanitarians because it permitted cruelty to animals. To the humanitarians, the building of elaborate structures like the wood rat's nest and the ability to adapt to new circumstances as did the glancing snake implied intelligence of a type similar to, if not equal to, that of humans. Bartram went so far as to grant higher morals to animals, in a certain respect casting them as the ideal citizens in the new American republic, as they "seldom transgress boundaries of Moderation and decency. & they often set us examples worthy of our imitation with respect to civil concerns" and do not engage in useless "National wars." This naturalist's breadth of experiences with wildlife allowed him to contribute new examples to the well-worn stories of cats, dogs, and long-suffering horses used in other eighteenth-century Anglo-American humanitarian literature and so promote the cause of kindness to animals.[47]

Wilson's field observations also provided a wealth of opportunities for new similarities between human and bird life beyond conventional linkages like love and dove. The bluebird epitomized romantic love through the offering of insect tidbits to its mate, and the cardinal who raised Wilson's cowbird appears an exemplar of parental devotion. A correspondent of Wilson, Dr. Gideon Smith, likewise read a lesson concerning the relationship of power and justice in the unusual text of the entrails of an eagle. As he reported to Wilson, who printed the letter, the presence of a fish in the intestines puzzled him until he learned of the eagle's propensity to take fish from ospreys: "Thus we see, throughout the whole empire of animal life, power is almost always in a state of hostility to justice; and of the Deity only can it truly be said, that justice is commensurate with *power!*"[48]

The naturalists further discerned in their fieldwork the centuries-old dream of the "peaceable kingdom," in which the separate species relax inherent animosities and coexist in harmony. Their figures of speech and general vocabulary applicable to humans imply, unsubtly, that mankind could similarly find peace. In perhaps the most famous passage of the *Travels*, Bartram relates that he saw, in the supernaturally clear waters of the Manatee Spring, fish in "the watery nations, in numerous bands," who would have been "at war" had they not been in the part of the stream that was "transparent as the air," and so speculates: "Do they agree on a truce, a suspension of hostilities? or by some secret divine influence, is desire taken away?" Wilson in his habitual bird-keeping once placed a bluejay, a notoriously quarrelsome species, into a female oriole's cage:

> She also put on airs of alarm . . . the Jay meanwhile sat mute
> and motionless in the bottom of the cage. . . . Accordingly in
> a few minutes, after displaying various threatening gestures
> (like some of those Indians we read of in their first interviews
> with the whites), she began to make her approaches, but with
> great circumspection, and readiness for retreat. . . . All this
> ceremonious jealousy vanished before evening; and they now
> roost together, feed, and play together, in perfect harmony
> and good humor. . . . This attachment shews that the disposi-
> tion of the Blue Jay may be humanized, and rendered suscep-
> tible of affectionate impressions, even for those birds which
> in a state of nature he would have no hesitation in making a
> meal of.[49]

The State Histories

Upon the arrival of nation and statehoods, inhabitants began to write their own "histories of place" with a renewed sense of pride, the best-known example being Thomas Jefferson's *Notes on the State of Virginia*, in which he refutes the belief of European naturalists, in particular the Comte de Buffon, in the degeneration of American species. Other writers of state histories inserted lists of plants and animals next to descriptions of climate, industries, agriculture, and the history of the state since its settlement. Incidents relating to individual species like the discovery by a Moses Catlin of Burlington of live frogs in the ground (purportedly discovered while digging a well in 1806) merited as much attention from the Reverend Samuel Williams in his *History of Vermont* as did human enterprises such as Revolutionary War battles.[50] Future writers of state descriptions would also include natural history data as a matter of course.

Between discussions of "political characters, genius, manners, employment, and diversions of the people," Jeremy Belknap's *History of New Hampshire* (1784–1792) includes chapters on "forest-trees and other vegetable production" and "native animals." A minister and founder of the Massachusetts Historical Society, Belknap characterizes his knowledge of natural history as "imperfect," yet he industriously compiled over 116 names of birds (with two naturalist friends, William Dandridge Peck and the Reverend Manasseh Cutler, supplying the binomial nomenclature). Although his descriptions are brief, he, like Jefferson, heartily enjoyed the opportunity to correct the despised Buffon and other European authors regarding their "incredible fables," such as those surrounding the beaver's ability to cut down huge trunks. He notes only those trees or flowers with "most remarkable qualities, whether salutary or noxious" to reduce the daunting number of plant species, and, like many of the authors of colonial histories, only briefly mentions broad categories of reptiles and insects, such as the "fly." As a matter of scientific record and perhaps of local pride, Belknap states his opinion that the black fly and cockroach were imported into New Hampshire. Natural objects associated with species became the equivalent of monuments, as in the case of Samuel Williams's careful recording of the "swallow trees" at Middlebury and Bridgeport, Vermont, in which large flocks would fly into a hole before sunset and roost. "A man who, for several years, lived within twenty rods of it" told Williams, "It was customary for persons in the vicinity to visit this tree, to observe the motions of these birds, and when any person disturbed their operations, by striking violently against the tree with their axes, the Swallows would rush out in millions, and with a great noise."[51]

John Drayton, during his tenure as governor of South Carolina, contributed more natural history lists among discussions of climate and human events and activities in his *A View of South-Carolina as Respects Her Natural and Civil Concerns* (1802). He bemoaned the "little leisure" with which he had to execute the work, yet he fully displayed the products of his state, "where the climate and soil of Carolina so powerfully assist the productions of the earth." The apparent scientific disinterestedness of natural history discourse made his claims of Carolina's natural wealth more credible.[52] Drayton pays the most attention to plants and often gave their scientific names, time of blos-

soming, and favorite soil—an understandable bias, as he was an
accomplished botanist who had compiled an unpublished botanical
catalog of the state. He dwells on the state's indigenous plants as opposed to imported "exotics," so that the readers would better appreciate the state's own "natural productions." Although he adds
little knowledge beyond repeating Bartram and Catesby in regard to
birds and animals, Drayton does offer the revealing observation that
"the buffaloe and cat-a-mount are entirely exterminated on the eastern side of our mountains; and the beaver is but rarely met with."
Natural history discourse had begun to act as a record for the disappearance of species; Belknap, for example, had remarked on the
dwindling numbers of deer and wild turkey.[53]

The natural history genre thus encompassed such philosophical
concepts as nationalism, geopolitical realities like statehood, and developments within the discipline itself, including the new emphasis
on life histories. The discipline and format, patterned so as to appear
restrictive, to the contrary enabled individuals (especially in the case
of Wilson) to achieve personal expression. Wilson, Abbot, and Bartram did not radically change natural history description and iconography, but they did enhance elements previously portrayed unevenly.
It is helpful to remember with these texts in which claims of total
originality are regularly professed that personal observations and ar-
ticulation are always based on existing structures. As Bartram had
built on Lawson and Catesby, and Wilson on Bartram, the next generation would follow and correct these predecessors. The inhabitants
now were using the genre of travelers.

2

Creating the Inhabitant Image: U.S. Natural History Monographs, 1825–1850

What may not be expected in a country like our own? Where the monstrous forms of superstition and authority, which tend to make ignorance perpetual, by setting bounds to the progress of the mind in its inquiry after physical truths, no longer bar the avenues of science; and where the liberal hand of nature has spread around us in rich profusion, the objects of our research?

Naturalist Richard Harlan in 1837[1]

Alexander Wilson left as a legacy the successful scientific monograph *American Ornithology*, whose physical appearance and original content matched or surpassed any other book so far produced in the United States. Between 1825 and 1850, the succeeding generation of naturalists seemingly divided up the animal kingdom among themselves and aimed to publish similarly beautiful works related to the United States. Botanists likewise favored expensively illustrated books portraying only American species. These naturalists' purpose, like that of Wilson, was to accentuate their research and "inhabitant" experiences and yet relate to the European natural history tradition. Several reached their goal while others' ambitions remain unfulfilled owing to uneven financial support and challenging publishing economics. Both the completed and incomplete works compel respect not only for the authors but also for the other Americans who contributed their efforts in the forms of observation and artwork.[2]

45

Creating
the Inhabitant
Image:
U.S. Natural
History
Monographs,
1825–1850

The European émigré Constantine Rafinesque predicted this trend in 1816: "The popular knowledge of the natural sciences has been prevented in the United States, by the first works published on them, having followed the model of the splendid European publications intended for the wealthy." More modestly priced textbooks, as chapter 4 discusses, were produced in the thousands from 1825 to 1850, so Rafinesque was not completely prescient. The publications of such nascent natural history societies as the Academy of Natural Sciences of Philadelphia, the New York Lyceum, and the Boston Society of Natural History also included pieces on American species. The quest to author major monographs like those "splendid European publications," however, transfixed most of the leading naturalists.[3]

A monograph could satisfy the continuing American desire to equal European productions in intellectual force and physical display. The illustrations in similar graphic techniques but practiced by Americans show new species, many specimens, and the characteristic habitats and gestures—opportunities denied to foreign scientists. The accompanying textual descriptions, particularly the newly expanded "life histories" format, allow for full explication of not only American discoveries but also the mistakes of the European authorities. The emphasis on original American material also minimized the substantial disadvantage of collections and libraries less comprehensive than those in Europe.

The general cultural ethos of the United States between 1825 and 1850 encouraged expansive ambitions celebrating American products, particularly natural ones. The nationalistic fervor of Alexander Wilson and his generation underwent another transformation, reemphasizing the uniqueness of the American land and the privileges of the persons inhabiting it. The merging of God and the unspoiled wilderness into a national quasi-religion in this era reinvigorated an enhanced awareness of and pride in the land's riches. The blossoming of landscape painting into an expression of admiration for God's handiwork is one important manifestation and justly explored in later Americanist scholarship. Naturalists' works also demonstrate an almost obsessive focus on native products, with Richard Harlan's patriotic exhortation which opens this chapter exemplifying these expectations.

This concentration on plants and animals fostered respect and, in many instances, empathy toward the subjects. Past writing about plants and animals without relating them to humankind had given implicit weight and autonomy to the subject; naturalists now explicitly wrote detailed "biographies" and heartily praised character traits. The intertwined similitude and otherness of the subject matter continued to attract naturalists just as birds' humanlike parental and romantic love and unique flight and anatomy had intrigued Wilson. Their close observation detected a decline in plant and wildlife populations. The monographs testify to these interests and concerns.

The Disciplines and Their Monographs

One useful approach to the monographs of the era is through their different disciplines—ornithology, mammalogy, entomology, malacology, herpetology, and botany. Natural history had developed from the broad discussions of all species to narrower topics devoted to smaller divisions. This specialization, beginning in the field of ornithology, for example, in the 1780s in Europe, stimulated the formation of elaborate classification schemas including rules for nomenclature beyond the Linnaean system embracing the plant and animal kingdoms.[4] Naturalists themselves in their observations and publications increasingly concentrated on a single field. The development of disciplines also fostered monographs devoted to specific subjects, such as Alexander Wilson's *American Ornithology*.

Discipline formation held special implications for American natural history. The concentration on native species throughout each discipline's monographs promoted the innovations suggested by *American Ornithology*. The authors generally included both physical and taxonomic descriptions and generously scaled life histories. In the case of fauna, accounts detailed mating, birth, favorite foods, and characteristic sounds; in the case of flora, they set forth the usual soil, climate, and life cycles of flowering and fruition. As in Wilson's case, naturalists' original research comprised intimate observation of species not only in the field but in the home, plus the contributions of other individuals who provided additional specimens or firsthand accounts of habits. The opportunity to make discoveries and name new species led Americans to participate in the nomenclature debates once owned by Europeans with their superior scholarship.

Some variations between the disciplines developed. Not surprisingly, ornithology, as represented in the works of John James Audubon and Thomas Nuttall, continued the life histories in prose related closely to that of Wilson. The prominent botanists, namely John Torrey and Asa Gray, confronted with a much greater number of species and genera including hundreds of specimens never published, favored brief lists and explanations of their classification. No monographs devoted to insects, reptiles and amphibians, and mollusks comparable to *American Ornithology* had yet been attempted, with the most complete descriptions scattered in European works, older works such as Catesby's *Natural History of Carolina,* and brief articles in the American scientific magazines recounting new species. Naturalists in the specialities of entomology, herpetology, and conchology openly and frequently deplored the past neglect of their subjects and hastened to give the particular subset of creation that fascinated them the prominence accorded to ornithology.

In terms of the visual imagery, ornithology, mammalogy, and the invertebrate specialities extended the iconography suggested by Wilson and John Abbot. Multiple specimens and varying backgrounds enriched the imagery of the living animal in its environs. Botanical illustration generally followed the more taxonomic format loaded with structural details, yet, in the hands of a few artists, implied the characteristic attitude of the living plant.

Ornithology

The surge of new editions of *American Ornithology* during the decade after its creator's death indicates British and American naturalists' respect for the breadth and accuracy of Wilson's work. George Ord, a prominent member of the Academy of Natural Sciences who had edited the ninth volume, reissued *American Ornithology* in 1825 with illustrations pulled from the original plates. Another edition was published in 1828. Charles Lucien Bonaparte, an expatriate nephew of Napolean and a naturalist specializing in the niceties of nomenclature, added a continuum of the work, with new American species, published between 1828 and 1833. A contingent of Philadelphia artists including Titian Peale and Alexander Rider (an artist originally from Germany, who had colored plates for Wilson) continued to delineate the birds in characteristic attitudes, and Alexander Lawson once again engraved. Bonaparte, who classified mostly according to external structures such as beaks, legs, and nostrils (as did most of his contemporaries), warned against careless depiction of the minute distinctions in these features. As he cautioned in his discussion of leg scales in hawks, not hesitating to critique Wilson himself:

We cannot let pass this opportunity of exhorting engravers, draftsmen, and all artists employed on works of natural history, never to depend on what they are accustomed to see, but in all cases to copy faithfully what they have under their eyes; otherwise, taking for granted what they ought not, they will inevitably fall into these gross errors. Even the accurate Wilson himself, or rather perhaps his engraver, has committed the same error in representing the feet of the swallow-tailed hawk. Of what consequence, will it perhaps be said in the form of the scales covering the foot of the hawk? But these afford precisely one of the best representative characters of groups, and it will, therefore, not be thought unnecessary to caution artists in this, and similar cases.

Theories of classification, reliance on empirical observation, and accurate transcription of such observation to the drawing, then to the plate, thus merged. The artist's and engraver's responsibilities were indeed onerous.[5]

In Wilson's native land, Robert Jameson, a professor at the University of Edinburgh, edited in 1831 a cheaper edition combining Wilson's and Bonaparte's works for a literary miscellany. Sir William Jardine, another Scottish naturalist, almost immediately edited Jameson's work (the manuscript copy demonstrates that Jardine literally "cut and pasted" Jameson's copy and added his commentary below) with more abundant footnotes on nomenclature derived from his own extensive knowledge of ornithological literature. Thomas Mayo Brewer, a Boston doctor, naturalist, and newspaper editor who later specialized in oology (the study of eggs), added a synopsis to the Jardine edition in his 1840 edition (later reissued in 1853 and 1854). Wilson, a proudly naturalized U.S. citizen, would not have appreciated the importation of a British version of his work.[6]

The success of *American Ornithology* and an actual encounter with its creator reshaped the career of an unsuccessful businessman who excelled in hunting and bird painting. John James Audubon had been collecting and drawing birds and animals since his teens in Nantes, France, and upon arrival in the United States at age eighteen in 1803, he further developed his own "style of drawing birds." To improve on his earlier attempts, which "made . . . some pretty fair sign boards for poulterers," he shaped birds threaded with wire into lifelike poses and traced their outlines onto paper divided into squares to make foreshortening easier. Concurrently, he developed a "mixed-media" approach to render more precisely the varied surfaces of feathers, beaks, and legs with pencil, pastel, ink, and watercolor. In 1811, as a merchant in Kentucky, Audubon met Wilson who was on a joint subscription and exploration trip and saw his "sale copy" of *American Ornithology*. An apocryphal tale relates that in Wilson's presence, Audubon's business partner whispered to him that his own drawings were superior.

49

Creating
the Inhabitant
Image:
U.S. Natural
History
Monographs,
1825–1850

After business failure had dampened his family's fortunes, Audubon and his wife, Lucy, decided that he should develop his ornithological talents with a view to an eventual publication like Wilson's. He built upon his portfolio of bird drawings—thanks in part to the first of his talented artistic assistants, Joseph Mason—with stays in Louisiana and Mississippi. In his first attempt to publish his work, he ventured to Philadelphia in 1824 only to offend Wilson stalwarts Alexander Lawson and George Ord, but he received enough encouragement from such respected naturalists and artists as Bonaparte and Thomas Sully to seek a British engraver. His family saved enough money to send him in 1826 to England, where his showmanlike personality and undeniably impressive portfolio created a sensation in the most influential literary and social circles and just as importantly convinced two proficient engravers, first William Lizars of Edinburgh and later Robert Havell Jr. of London, to undertake his project.

As in Wilson's project, subscribers' payments financed production. The enormous price of ten dollars or two British guineas per part, culminating in four volumes for a total of 435 plates, was needed to pay the costs of huge copper plates (in a size generally reserved for maps), Havell's services in combining etching, engraving, and another method of printmaking, aquatint, used to translate Audubon's watercolor drawings, and the colorist's labors. It also covered Audubon's own living expenses, including trips to the United States to search for birds and, as he joked, the rarest birds of all, subscribers. The process of creating and overseeing *The Birds of America* absorbed the next twelve years of the Audubon family's lives.[7]

Audubon and his artistic collaborators synthesized elements of the previous tradition into a distinctive style advancing "the living bird in its environment." He used life-size images, as Wilson and other Europeans had on occasion, but took the practice to an extreme in representing full-size even the large water birds and birds of prey. Wilson and others had striven to render lifelike postures, but Audubon more consistently employed striking foreshortening and a much greater number of figures. Unlike previous illustrators, who had often omitted informative scenery, he always incorporated accompanying backgrounds, for smaller birds like the warblers, branches or flowers silhouetted against a white background, and for larger birds like the herons, a full-scale landscape.

The various poses and accessories created images of individuated species engaged in typical actions. The compositions most discussed by contemporaries and later critics were the elaborate dramas like the famed mockingbird scene in which parents fight off a rattlesnake in a battle of good versus evil. Far more prevalent, however, are the compositions designed to feature the bird's unique character, such as the scene in which the Carolina parakeets congregate (Illustration 2.1). The many scenes of nesting and feeding young relate to the dual human and avian experience of raising offspring.[8]

His study of wildlife, in- and out-of-doors, fed Audubon's innovations. Like Wilson, he kept birds as pets and, in the case of the least bittern, carefully noted its striking extensions of form which he imitated in setting up the freshly killed bird. He claimed that his observations validated the outré poses of birds wildly writhing in flight, and, as with Wilson, contributed to his knowledge of the correct accessories of nests, food, and general habitat. The designs, seemingly embodying freedom of movement, are in fact constructed to display every distinguishing part as required in ornithological illustration. His range of experiences and years of collecting a variety of eggs, chicks, and male and female specimens made multiple-figured compositions possible. One critic discerned the mesh of experience and composition in his style: "half the natural history of the bird may be read in these graphic delineations."[9]

The new and extended use of varying viewpoints underscores his experience and empathy with his subjects. In other natural history illustration, the relationship between viewer and subject in terms of sight line is indeterminate, although the subject dominates the space. In Audubon plates, often the viewer sees the birds at ground level as did Audubon, the hunter stalking his prey. Many scenes are "ornithocentric" (to borrow a term from an astute critic of Audubon, Robert Henry Welker), as the viewer is transported to the branches (as with the Carolina parakeets) or on the water to observe birds literally on their own territory. The birds at times are aware or wary of the viewer's presence, as in the multiple glances in the parakeet plate, but often are oblivious to the human gaze as they enact their own lives.[10]

The format of natural history texts required Audubon to write technical descriptions to accompany his plates, so he hired a rising young Scottish naturalist, William MacGillivray, to help with *The Birds of America*'s accompaniment, the five-volume *Ornithological Biography* (1831–1839). MacGillivray, himself an ardent "field" naturalist who ventured into the cold in order to check birds' stomachs to research winter feeding habits, compiled the synonymy and physical descriptions (complete with detailed analyses of gastrointestinal tracts) and

Illus. 2.1. "The Carolina Parakeet." Hand-colored engraving by Robert Havell Jr. after John James Audubon, for Audubon, *Birds of America,* vol. 1, Plate 26. (Gift of Mrs. Walter B. James. © 1997 Board of Trustees, National Gallery of Art, Washington.)

sensitively edited Audubon's voluminous life history segments for each species. The long life history portions relate apparently every example of Audubon's and his acquaintances' encounters with the species and often correct past historians' portrayals by using his superior specimens, plates, and observations. He asserts that his "field" experience uncovers traits distinguishing species unknown to "cabinet" or "closet" naturalists who work only from dead specimens: "But in the most intimately allied species there are always marked differences in habits, and especially in the sound of the voice."[11]

"Intensive" reading had again transmitted the natural history genre. Although his text corrects Wilson publicly (his private marginalia on his copies of *American Ornithology* more pungently note his differences on habits, as in his exclamation, "All my eye and Betty Martin," near Wilson's describtion of the wood ibis as "solitary"), Audubon constructed his work on the scaffold of Wilson's just as Wilson had based his lists and nomenclature on other authors. His constant rereading of that naturalist's singular prose style unwittingly shaped his own individual voice. He assumed Wilson's ability to mold personal visual and aural observation into the generalized description of an entire species. A contemporary wrote of this purple grackle passage, "Observe how, in the following sketch, every thing is exhibited to the eye":

> No sooner has the cotton or corn planter begun to turn his land into brown furrows, than the crow-blackbirds are seen sailing down from the skirts of the woods, alighting in the fields, and following his track along the ridges of newly-turned earth, with an elegant and elevated step, which shows them to be as fearless and free as the air through which they wing their way . . . the bird stops, spreads its tail, lowers its wings, and, with swelled throat and open bill, sounds a call to those which may chance to be passing near. The stately step is resumed. Its keen eye, busily engaged on either side, is immediately attracted by a grub, hastening to hide itself from the sudden exposure made by the plough. In vain does it hurry, for the grakle [*sic*] has seen and marked it for its own, and it is snatched up and swallowed in a moment.[12]

The author goes beyond scientific terminology and chooses adjectives with human connotations like *elegant* and *keen* to portray the grackle's movements. Sometimes he and other naturalists overtly draw comparisons between human and avian actions, but often the texts employ this language to convey more precisely the unique characteristic or "otherness" of the subject.

The life histories alternate general descriptions with intensely personal interjections of specific remembrance. Often exhorting his "dear reader" to share his feelings of wonder, Audubon launches into rhapsodies:

> Could you, kind reader, cast a momentary glance on the nest of the Hummingbird, and see, as I have seen, the newly-hatched pair of young, little larger than humble-bees [*sic*], naked, blind, and so feeble as scarcely to be able to raise their little bill to receive food from the parents; and could see those parents, full of anxiety and fear, passing and repassing within a few inches of your face, alighting on a twig not more than a yard from your body, waiting the result of your unwelcome visit in a state of the utmost despair,—you could not fail to be impressed with the deepest pangs which parental affection feels on the unexpected death of a cherished child. Then how pleasing is it, on your leaving the spot, to see the returning hope of the parents, when, after examining the nest, they find their nurslings untouched.

Here, the author's extraordinarily close physical presence fosters not only sympathy for the birds and the overt comparison with human actions but the keen desire to impart his message of noninterference (somewhat ironic considering his intrusion on the scene) to the reader.[13]

An Englishman who emigrated to the United States in 1808 to pursue his natural history interests, Thomas Nuttall was best known for travel accounts and botanical books at the time he published his *Manual of Ornithology* in 1832 (the second volume, *Water Birds*, was published in 1834). His introduction, while praising Wilson and Audubon, tactfully notes the need for a less expensive work, and indeed, he skillfully abridged those author's prolix descriptions and added wood engravings adapted from their engraved plates. Nuttall, however, contributed his own phonetic imitations of bird song (the purple grackle's cry is "crick crick cree" and the less familiar call of the chestnut-sided warbler " 'tsh 'tsh 'tsh 'tshyia") and was able to write as "graphic" original description as his predecessors. With unexpected similes, he conveys the speed and sound of the Nootka hummingbird he saw on a western expedition:

Towards the close of May, the females were sitting, at which time the males were uncommonly quarrelsome and vigilant, darting out at me as I approached the tree probably near the nest, looking like a vivid coal of fire, passing within a very little of my face, returning several times to the attack, sinking and darting with the utmost velocity, at the same time uttering a curious reverberating sharp bleat, somewhat similar to the quivering twang of a dead twig, yet also so much like the real bleat of some small quadruped, that for some time I searched the ground, in place of the air, for the actor in the scene.[14]

The reader also looks, so vivid is the description.

Mammalogy

Wilson had poignantly dreamed about a version of American mammals as a counterpart of his *Ornithology;* Audubon, then his sons Victor Gifford and John Woodhouse, in the 1840s and 1850s produced their *Quadrupeds of North America* as a successor to the *Birds.* Two decades earlier, two Philadelphia naturalists had come to verbal blows over their versions of this portion of the animal kingdom. Richard Harlan and John Godman almost simultaneously published *Fauna Americana* and *American Natural History,* respectively (Harlan's work was published in 1825, Godman's three volumes between 1826 and 1828). Godman accused Harlan of plagiarizing the Frenchman Desmarest's work, while Harlan published an open letter refuting the charge. Godman's work became the better known, through more extensive distribution and subsequent editions, inclusion of engraved illustrations, and more accessible prose.[15]

Godman, a practicing physician, blended his considerable knowledge of the Frenchman Georges Cuvier's standard work on classification, *Le Regne Animale,* with his own dissections to expound lengthy physical descriptions. His life histories of species, which he called "general histories," most clearly reveal his pride in the firsthand observations that "delayed appearance of the work . . . in order to procure certain animals, to observe their habits in captivity, or to make daily visits to the woods and fields for the sake of witnessing their actions." He rejoices in diminishing with his superior observations such popular exaggerations as the wolverine's gluttony, but he also asserts unpleasant truths like that of the raccoon's "mingled expression of sagacity and innocence" masking its "bloodthirsty and vindicative" spirit in killing chickens "without consuming any part of them,

55

Creating
the Inhabitant
Image:
U.S. Natural
History
Monographs,
1825–1850

except the head." Godman appears to have witnessed many aspects of common mammals' lives, including the raccoon's unusual coition, the description of which he presents in Latin for the benefit only of naturalists and other professionals.[16]

His general history of the shrew mole exemplifies Godman's passion for original observations. He dug into the ground to measure galleries excavated by moles in locations varying from those "stated in books" and actually broke the tunnels in order to observe whether the moles would repair them. Remarks such as "when the animal is finally dragged from his retreat, he frequently inflicts a severe bite on his disturber" bespeak Godman's personal sacrifices. His dissection of the mole's shoulder muscles and narrowing pelvis demonstrated that the interior anatomy complements the exterior shape of the paws and the smooth fur. Such analysis of the adaptation of the shrew mole's physique to its lifestyle of burrowing reinvigorated the natural theology tradition of praising God's creative perfection: "Every circumstance seems to be studied in the shrew mole with a view to facilitate its progression under the surface of the earth." Although the shrew mole at worst seems a pest who eats grass roots (Godman tells farmers that the mole also destroys unwelcome grubs) and at best more closely resembles to the ordinary viewer "a small stuffed sack than a living animal," under his investigation the species becomes highly interesting and beautiful in its adaptation. Godman became so identified with the species that an obituary said of him, "in investigating the habits of the shrew mole, he walked many hundred miles."[17]

Godman described only the major species and left many untouched, as a budding and yet unknown naturalist, Spencer Fullerton Baird, complained to Audubon in 1840, when Audubon had already embarked on his production of lithographed plates in magnificent imperial folio size. Audubon set the animals, often in herds or family groups, into landscapes of desert, prairie, and meadows, and smaller pastoral scenes provided by his son Victor Gifford, who had trained as a landscape painter. His lithographer, John T. Bowen, an Englishman transplanted to Philadelphia, probably employed "engraving on stone," the technique in which the traditional engraver's tools make very fine lines and details into the ground covering the lithographic stone. An oily fluid is pressed into these "engraved" lines, and then water washes away the ground (the design made by the lines remains because of the antipathy between the oil and water). The ink adheres to the design, allowing printing to occur. Such a practice, because it allowed for greater precision than ordinary lithography, suited well the details of teeth and claws important to the then-current systems of classification and also contributed to the hair-by-hair texture of the fur (Illustration 2.2). John Woodhouse Audubon, himself an accomplished artist in the animal painting tradition, completed one-half of the designs after his father's mental health declined in 1848.[18]

John James Audubon was fortunate in gaining the services of Charleston clergyman and naturalist John Bachman. Bachman's close friendship and support of Audubon had informed the latter's *Ornithological Biography,* replete with frequent references to Bachman's bird observations. Bachman himself was confident enough in his expertise to "row you [Audubon] & Wilson a little up salt river in order that I may give a regular but genteel ducking" in his article on "the moulting of bird and shedding of the hair in animals." Bachman, in supervising the *Quadrupeds* texts, edited Audubon's notes of personal observations and added his voluminous commentary on nomenclature and habits. Bachman's thoroughness is evident in his text subdivisions: synonyms, physical appearance, color and dimension of specimens, geographical distribution, and habits. His correspon-

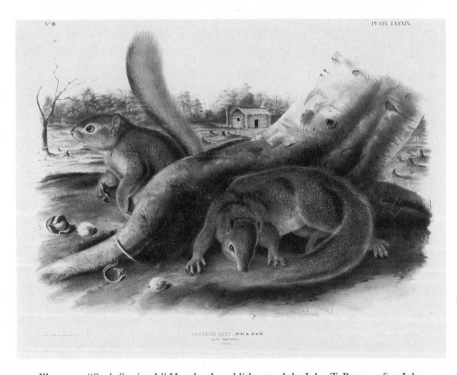

Illus. 2.2. "Say's Squirrel." Hand-colored lithograph by John T. Bowen after John James Audubon, for Audubon and Bachman, *Viviparous Quadrupeds of North America,* Plate 89. (Courtesy of the Department of Rare Books and Special Collections, Princeton University Library.)

dence to the Audubons bespeaks his determination to present the most thorough and accurate descriptions published by Europeans or Americans: "This studying every tint and hair and going over hundreds of old authors is not fun—It must however be done to settle the species, . . . you will learn bye and bye that I have not only been a student of books but of nature."[19]

Bachman's "books and nature" were to provide the structure for his research and subsequent text. He struggled to identify the many species and varieties of squirrels by comparing color, teeth, geographic range, and habits of whatever specimens he could check. Audubon jokingly admitted that the subscribers, who perhaps wanted more variety in the imagery, would not immediately appreciate the growing number of similar species but, "although you should find that more Squirrels inhabit our forests than you expected or desired to be figured in this work, we assure you it would give us pleasure to discover a new species at any time!" His belief in common sense, similar to that of Wilson, led Bachman to dismiss robustly such "romantic stories" of European travelers as those that attributed unusual powers of intelligence to the beaver, which, in his view, behaved stupidly apart from its ability to build dams. Bachman's research into the gestation of the opossum, begun in his youth, vividly conveys his eagerness to clarify what had mystified other naturalists "unusually long." European and American observers could not determine whether the young first grew in the uterus and, if so, how they got from the uterus into the pouch. Bachman, in a number of attempts to capture pregnant females, offered premiums to neighboring servants but instead at one time received thirty-five males. Luckily, a week later friends brought him the proper sex, in which through dissection he observed the young in utero. Once, "without immediately knowing," he observed the female that he had brought into his study lick her tiny young into the pouch and so watched "the manner in which the young at birth reach the pouch," ending the speculation of an interior channel between the pouch and uterus.[20]

To Bachman as to Godman, the inspection of mammals continually revealed new details and intricate structures. He acknowledges "the birds, with their varied and pleasing forms, their gay and beautiful plumage, their tuneful throats, and their graceful movements through the air" present showier attractions to the student of nature than quadrupeds, because most mammals are active nocturnally and are visually duller than birds. Bachman assures readers that once the specimens are located and their appearance is given further inspection, "we find in them matter to interest us greatly, and arouse our

57

Creating
the Inhabitant
Image:
U.S. Natural
History
Monographs,
1825–1850

curiosity and astonishment." The intricate patterns, mimetic colors, and glorious textures of the Audubon plates bring out the visual wonders, while Bachman's text sets forth the fascinating habits.[21]

Entomology

While mammals may have held less interest than birds, insects generally appealed even less than mammals. Thomas Say, a Philadelphian whose familial relationship with William Bartram nurtured a youthful interest in natural history and who was one of the founders of the Academy of Natural Sciences in Philadelphia, sought to elevate their status in the vehicle of a national monograph. He admits in his prospectus to *American Entomology* (1817, 1824–1828) that ornithology and botany had surpassed entomology in terms of "indigenous" works in the United States because of the lurking suspicion that "insects are despicable because minute." Although the butterflies and bees look "captivating to the beholder . . . the variety of systems, the obscurity of the distinctive characteristics, and often the great requisite nicety of discrimination upon which some of those systems are founded" deter potential students. Say's own correspondence to the very few naturalists specializing in entomology in the first half of the century demonstrates the difficulties of determining whether the description of a specimen had already been published and of naming a new species according to the established classification systems because of the limited number of illustrated European works available. He apologized to an older colleague, John F. Melsheimer, that the names in his insect collection "may be wrong as I have not the chance of reference to many books." The situation continued to look bleak, for "our booksellers [are] unwilling to incur the risk of importing costly and splendid works." Say soldiered on, "acting the part of a pioneer in an untried path . . . destitute of a colleague."[22]

Particularly distressing to Say were Europeans' superior knowledge and more numerous publications concerning American insects. Throughout his life, as his most recent biographer, Patricia Tyson Stroud, has stressed, he championed Americans publishing studies of American insects. He therefore excoriated his contemporary John E. LeConte for giving his superior collection of beetles to the French entomologist Dejean and arranging for his butterfly specimens and drawings to be published by the Frenchman Boisduval. Say urged a fellow American naturalist, Thaddeus Harris, to prevent further depredations by publishing: "Do then, if agreeable to you, give us a paper descriptive of the unknown insects of this beautiful & neglected Order, & let them not be wrested from their own countrymen, thus unnaturally."[23]

Say himself intended his publications to contend on the world stage. He designed his *Entomology* to equal European standards and even acknowledged Edward Donovan's *History of British Insects* as a model in text and plate format. There were far too many individual insect species to devote entire plates and segments to each one, as was often done for birds and mammals, so, like Donovan, he gave the genus's characteristics followed by individual descriptions and observations and arranged for several members of the same genus to be figured on each plate. The letterpress printing was of the highest quality available in the United States, and Say wished "the plates may be well done, and faithfully colored, as to present accurate and characteristic representations of the originals."[24]

59

Creating
the Inhabitant
Image:
U.S. Natural
History
Monographs,
1825–1850

Taxonomic concerns, reflected mostly in the difficulties in providing accurate descriptions following the family and genera guidelines in often conflicting European sources and the process of naming new species, occupied much of Say's energy and hence his text. However, his descriptions vividly relate the "character" and habitual movements of insects that he and others observed, much like the other disciplines' life histories. The Smerinthus moth flew in its distinct manner; "unlike the sphinx [moth] and Humming-bird, their flight is heavy and reluctant, and they receive food only in the state of repose." Members of the *Cryptocephalus* genus (a beetle) "on the approach of danger, . . . counterfeit death by retracting the feet and antennae close to the body, and permit [themselves] to fall from any height to the ground." He had participated in the Long Expedition crossing the Rocky Mountains in 1819, so his remembrances flavor his habitat description of the *Lytta albida:* "It appeared to be feeding upon the scanty grass, in a situation from which the eye could not rest upon a tree, or even a humble shrub, throughout the entire range of its vision, to interrupt the uniformity of a far outspreading, gently undulated surface, that, like the ocean, presented an equal horizon in every direction." He explained that the ichneumon larvae, growing inside the larvae of moths and butterflies, ate the soft tissues before the vital organs and so extended the lives of their food source. Bees and wasps had fascinated humans with their intricate nest architecture since ancient times, and later eighteenth-century naturalists like René de Reaumur published life histories based on observations of the process of colony-building. Say echoes this tradition when he wonders aloud that complexities of wasp nest-building confound the difference in his mind between instinct and reason: such species "exhibit such eminent proofs of intelligence."[25]

Designs were drawn by Titian Ramsay Peale, scion of the famed art and natural history family; Charles LeSueur, a transplanted

French naturalist who had published drawings in some of the French government's magnificent expedition folios; and William Wood, a Philadelphia artist and naturalist. Cornelius Tiebout, who had previously distinguished himself as a portrait engraver, used his skills in the stipple technique and etching to capture the precise delineations of joints, antennae, and markings. In many plates the species is isolated against a white background in a symmetrical format to stress the physical differences otherwise difficult to grasp visually. Others, however, show the species interacting with the environment. The insects crawl up stems and onto leaves (Illustration 2.3), while butterflies rest on branches. The ambitious work limped to a close in 1828 after completion of fifty-four plates when Say moved to the New Harmony community in Indiana and his publisher Samuel Mitchell in Philadelphia refused to continue the financially unprofitable project.

Conchology and Malacology

Despite his move to Indiana and the termination of his entomology book, Say continued his pioneering American publications with his projected *American Conchology*. Ironically, many Americans were familiar with the appearance of imported shells prized for their beauty and used for household ornaments but not with the natural history of foreign or native species. Already the author of numerous articles on American shells, Say wanted to publish another monograph that portrayed all native species, although he was isolated from books and other naturalists in his new home. While advertising for subscribers in the *Disseminator*, the colony's newspaper, he personally oversaw production of *American Conchology* on New Harmony's community presses. The more than one hundred subscribers were centered in the Midwest. His wife, Lucy, undertook the delicate task of drawing the shells, and Cornelius Tiebout and two others engraved the sixty-eight plates. Lucy and two Tiebout children hand-colored the thousands of copies. Say was limited to representing shells sent to him by his brother in New Jersey, some Academy of Natural Sciences friends, colleagues in Charleston, and those he himself had collected. Say's physical decline and eventual death in 1834 stopped the publication begun in 1830 at part six, with the seventh part issued several year afterwards.[26]

Two men, Amos Binney and Samuel Stehman Haldeman, took up Say's task of producing national monographs on conchology. A wealthy Boston merchant who studied shells to relieve his mind "from the all-absorbing cares which are incidental to mercantile pursuits," Binney died in 1847 before his study on U.S. land snails could

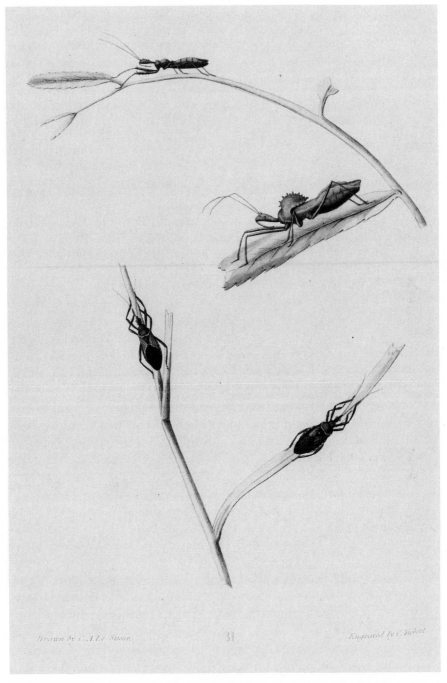

Illus. 2.3. *"Redivius."* Hand-colored engraving by Cornelius Tiebout after Charles LeSueur, for Say, *American Entomology,* vol. 2, Plate 31. (Courtesy of the Library, Academy of Natural Sciences, Philadelphia.)

be published. Augustus A. Gould, another Boston conchologist, edited the first three volumes of *The Terrestrial Air-Breathing Mollusks of the United States,* published between 1851 and 1857. Binney declared that he would restore American species inaccurately described by Europeans to America, and Gould promised the "completeness and the style of its mechanical execution should be unsurpassed by any similar work." S. S. Haldeman, who as a young man had "no taste for his father's saw mill business," according to a later biographer, "hated feeling tramelled" and enjoyed the outdoors. He learned taxidermy from a preacher near his home in central Pennsylvania and more formally studied natural history at a series of lectures at the University of Pennsylvania. After serving on a state geological survey in 1836, he began to focus his study of freshwater shells stimulated by boyhood memories of scavenging on the Susquehanna. He wrote and oversaw publication of *A Monograph of the Freshwater Univalve Mollusca of the United States* between 1840 and 1845.[27]

The science was slowly transforming in Europe into malacology, the study of the living animal within the shell. Binney and Haldeman noted that age and climatic differences produce variations in shells of a single species, which had led Europeans, who had access to fewer shells, to believe the shells represented different species. Gould boasted that Binney had "kept most of the species in captivity for months, that he might observe their habits, the variations they exhibited, and the changes they underwent by age, food, etc." Haldeman warned of the difficulties the "extent of variation and its possible limit, if there is such a limit in nature" posed in positing species and reminded his readers that, in his subject area particularly, "many species are less distinct in nature than they are represented to be in books." Conchology formerly had depended solely on taxonomy of shells, but both men agreed that the animals' anatomy often indicated genus and species and observed the habits of the animals, especially to see whether they respired on land or in water: "Every naturalist should therefore test his opinions by an intimate acquaintance with living species existing in their natural conditions." In a footnote, Haldeman discreetly boasted of his field research beyond the homefront: "In the summer of 1841, accompanied by an assistant . . . I drove a single horse . . . the distance of 1800 miles in 45 consecutive days, and often over mountainous regions and the most execrable roads, frequently breaking shafts, springs, and other portions of the vehicle, and examining every stream upon the route."[28]

The plates' iconography subtly reflects the shifts in the field toward studying the living animal and its geographical distribution.

63

Creating
the Inhabitant
Image:
U.S. Natural
History
Monographs,
1825–1850

Lucy Say's images for *American Conchology* recorded only one example of each species' shell in its interior, exterior, and side views. Binney's and Haldeman's works place shells in various stages of growth and from different geographic locations on one plate to demonstrate variation, as in the nine specimens of *Limnea fragilis*. In some instances, the living animals crawl in their shells over the empty background (Illustration 2.4, middle, top, and bottom). Alexander Lawson, still the dean of Philadelphia natural history engravers, engraved the majority of plates in both works, with his daughter Helen and son Oscar drawing many of the specimens. The engraver's lines imitated the lines of the drawings, which, in turn, had mimicked the whorls and scale of actual shells (in one image, the crack of the original specimen appears). Helen Lawson colored most of the plates. Her precise handling of highlights and translucent colors within a narrow range of browns and greens makes each shell gemlike.[29]

Herpetology

Because of their potentially dangerous nature and inhospitable habitat, snakes and amphibians were seldom described. John Edwards Holbrook, the preeminent American herpetologist in the first half of the century, who authored *North American Herpetology*, almost glories in the difficulties presented in his state of South Carolina: "Inhabiting, for the most part, deep and extensive swamps, infected with malaria, and abounding with diseases during the summer months, when Reptiles are most numerous, time is wanting to observe their modes of life with any prospect of success. Regarded, moreover, by most persons as objects of detestation, represented as venomous, and possessed of the most noxious properties, few have been hardy enough to study their character and habits." To Holbrook, however, who had studied briefly with French invertebrate specialists in the 1820s before resuming a busy medical career, the amphibians and snakes offered a multitude of interesting adaptions to varying modes of life—organizations "for creeping, others for walking, for swimming, and even for flying" of potentially endless interest.[30]

The few herpetological works before the 1840s had either been unillustrated or confined to state catalogs; Holbrook was the first to match original illustrations with extended text in an attempt to span the land. The first volume appeared in 1836, but, owing in large part to Holbrook's search for the best information and specimens, was revised in a five-volume second edition in 1842. Like his fellow Charleston naturalist John Bachman, he enjoyed correcting other naturalists' miscues, even delving into the eighteenth-century litera-

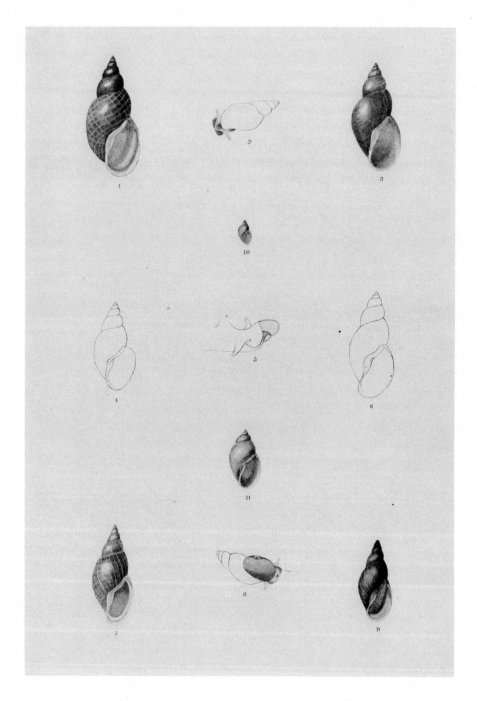

Illus. 2.4. *"Limnea."* Engraving by Alexander Lawson after Helen Lawson and colored by Helen Lawson, for Haldeman, *A Monograph of the Freshwater Univalve Mollusca,* part 2, Plate 6. (Courtesy of the Library, Academy of Natural Sciences, Philadelphia.)

ture to correct Catesby and Linnaeus. Holbrook's greater access to specimens across the country and exact knowledge of their sources helped him to avoid the European authors' errors in geographical distribution and physical description, and he boasted that, because specimens were fresh and not bottled in alcohol, their colors had not faded nor their features become less distinct, thus much improving his classifications.

65

Creating
the Inhabitant
Image:
U.S. Natural
History
Monographs,
1825–1850

The "habits" of his descriptions demonstrate his years of close observations in- and out-of-doors. He captured and kept live specimens for artists to draw and for him to experiment with, once placing nonvenomous snakes in the same cage with the poisonous water moccasin to watch them "all evinc[ing] the greatest distress, hanging to the sides of the cage, and endeavouring by every means to escape from their enemy, who attacked them all in turn." Holbrook, evidently well supplied with the species, placed two other water moccasins in the cage, which calmed the rampant individual. He poked and prodded the nonvenomous *Heterdon platirhinos* snake "many times, yet I have never seen it bite or lay hold of any object offered it." Although reptiles were held traditionally to be especially removed from human characteristics, Holbrook's interactions with various species allowed him to grant them behavioral traits. One water snake "was a bold animal, even in confinement, one of the few snakes that will, in such a situation, readily devour its prey." A species of frog "is timid and remarkably gentle in its habits, remaining concealed during the day in some dark place, and only venturing out as the dusk of evening approaches." His outdoor study led him to emphasize snakes' utility in keeping down the rat population in rice fields, in order to quell "the universal prejudice against the serpent tribe caus-[ing] its destruction from all hands."[31]

To complement his painstaking original research, Holbrook invested much time and energy in the production of plates. By the second edition, over sixteen artists had drawn for *Herpetology*, including John H. Richard, a Philadelphia artist who later illustrated many federal government natural history surveys, and Maria Martin, a fellow Charleston resident and Bachman's sister-in-law, then wife, who was more accustomed to drawing background plants for Audubon than facing live snakes. Professional artists proficient in other genres, such as Albert Newsam, the deaf-mute known for his portraits, drew for Holbrook. The first Episcopalian bishop of Vermont, the Reverend John Henry Hopkins, demonstrated considerable ability in his original salamander drawing, with its curving tail and display of claws, sent in lieu of a specimen (Illustration 2.5). All these artists imparted a sense of life and motion (in Holbrook's words, "spiritedness"), although the animal was separated from any environment. The snakes are coiled writhingly with the scales on their undersides cleverly and consistently exposed, while the dorsal and ventral aspects are given for the amphibians. The Philadelphia firm of Peter Duval experimented with a form of color printing, lithotinting, to register color more consistently. Holbrook and a friend in Philadelphia, Edward Hallowell, checked proofs to correct color, and, in the most concerted attempt to maintain the highest standards of accuracy, Holbrook "recalled" the first edition from subscribers to replace plates and text with new, improved images and information.[32]

Botany

Botanists knew they could never illustrate each of the thousands of species native to the United States, so the earlier American productions such as Jacob Bigelow's *Medical Botany* and Benjamin Smith Barton's *Elements of Botany* (1803) offered few plates compared to the

Illus. 2.5. *"Menobranchus maculatus."* Lithotint (?) by P. S. Duval after John Henry Hopkins, for Holbrook, *North American Herpetology,* vol. 5, Plate 37. (Courtesy of the Library, Academy of Natural Sciences, Philadelphia.)

number of species they verbally described. Despite such relative frugality, botany still had been the best illustrated of all disciplines in the early republic. This was due to the many European publications interested in American products for agricultural and gardening purposes as well as the cheaply produced herbals with woodcuts stemming from the Germanic tradition. In the 1830s, John Torrey began the mammoth task of describing and cataloging all indigenous species according to the "natural system" of classification, which gradually supplanted the "artificial" or Linnaean system in the United States. However, his *Flora of North America* included no plates. The natural system placed a premium on understanding many parts of the plant for its classification, compared to the Linnaean system in which only the pistils and stamens determined the class.

It was his one-time protégé and then coauthor of the *Flora*, Asa Gray, who commenced the major botanical project with original plates the star attraction. Gray's previous work with Isaac Sprague, a young artist near Gray's home base at Harvard College, and Joseph Prestele, a German emigrant lithographer living in a Christian community in Ebenezer, New York, inspired him to produce images of American species according to a European model: "By the aid of my young and excellent artist Sprague's drawings, and Prestele to engrave cheaply and neatly on stone, I am going to commence a Genera Illustrata of the United States, like T. Nees von Esenbeck's 'Genera Germanica Iconibus Illustrata.' " Such works featured one member of a genus selected often for singular beauty.[33]

Gray knew his colleagues could produce such a work. Sprague as a youth in Hingham, Massachusetts, had been inspired to draw birds after reading Thomas Nuttall's *Manual of Ornithology* (his drawings so impressed Audubon that Sprague served as an artistic assistant on the 1843 expedition to the Missouri River). A Harvard colleague in 1844 recommended Sprague to Gray, who needed an illustrator for his many projects. The relationship lasted twenty years. Gray taught Sprague how to dissect plants and use the microscope to understand and render anatomical details of fruits and stems; Sprague became so adept that he "discovered some new quiddities about the position of the ovule in Ranunculaceae." Prestele could equal Europeans' handiwork because he was in fact a German-trained botanical artist. He had studied botanical painting and its closely related genre, flower painting, since adolescence and learned lithography in Munich, the birthplace of the technique. Having worked as a farmer in his religious community after emigrating in 1843, he fortuitously wrote to Gray, "tender[ing] you my services whenever you will afford

67

Creating
the Inhabitant
Image:
U.S. Natural
History
Monographs,
1825–1850

me an opportunity to render any." His role involved hand-coloring and "engraving on stone," the special lithographic technique. The distance between Gray in Cambridge and Prestele in New York state, then Amana, Iowa, necessitated lengthy delays in correcting proofs and coloring jobs. Moreover, Prestele had to obtain his heavy stones from the New York City firm of Endicott & Co. and then usually send back the finished stone for them to print the plates.[34]

The plates of the *Genera Florae Americae Boreali-Orientalis Illustrata* (1848) blend the traditional botanical conventions of showing the plant uprooted against an empty background and the new importance placed by the natural system on details of more anatomical structures. The plants came dried from collectors over the United States or were grown from donated seeds in the Harvard Botanical Gardens. Sprague consistently conveyed a vital quality to his plants despite their artificial setting, perhaps because he saw many actually live through life cycles of bloom and fruition. For example, in addition to the many details of dissected parts, he gave the characteristic attitude of the springing violets (Illustration 2.6). Prestele's botanical background helped him translate the anatomical details and varying tactile textures with respect. Sprague spent so much time over his dissections and drawings, however, that Gray complained Sprague "barely made day wages." With so much time and effort required to produce the plates (and so little profit to show for it), Gray concluded the *Genera* after only two volumes, well short of the projected ten. In a letter to a colleague, he blamed the incomplete project on Sprague's slowness—"yet I could bring out vol. after vol. of Genera Illustrata . . . if Sprague could only accomplish the drawings"—rather unfairly, as his short catalog descriptions would naturally take less time than drawing and executing the elegant plates.[35]

Themes across the Disciplines

These authors claimed that their own particular discipline and its subjects held special pleasures which only increased with study and observation. Indeed, the advancements in knowledge of their objects of attention did not lessen the individuals' inquisitiveness but formed a foundation for future research. For example, Audubon and Nuttall desired further knowledge of the species hitherto described by Wilson, if only to correct his text and add their personal observations. The more live snakes of a particular species Holbrook could collect and watch over time, the better. Their subsequent publications would enable others to study the discipline more effectively—no more the plaints for books like those from the young Thomas Say.

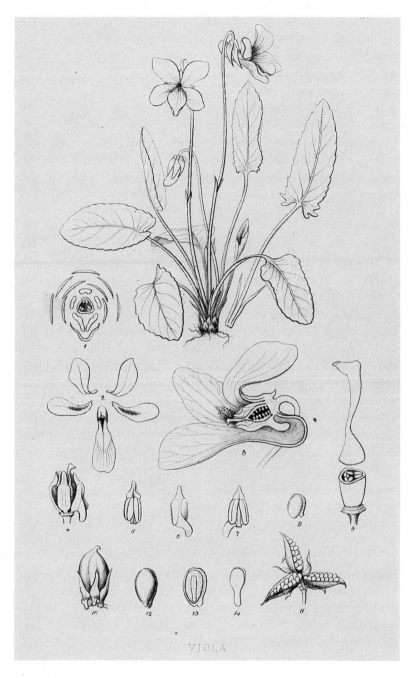

Illus. 2.6. "Viola." Lithograph (engraving on stone) by Joseph Prestele after Isaac Sprague, for Asa Gray, *Genera Florae Americae Boreali-Orientalis Illustrata* (Boston: James Munro, 1848), vol. 1, Plate 80. (Courtesy of the Library, Academy of Natural Sciences, Philadelphia.)

Despite their promotion of their specialties, the naturalists worked with the same prose and visual conventions and shared a common outlook upon the natural world. Naturalists often relied on and featured prominently in their monographs specimens and observations from individuals who did not themselves actively publish but who were in some cases as dedicated to the disciplines as the authors. Such intensive and original research resulted in several publications, such as the Audubon folios, Holbrook's *Herpetology,* and Gray's *Genera,* that ranked with those of the Old World in terms of scholarship and presentation despite the scarcity of resources for producing elaborate works. The more the naturalists examined their homeland, the more they realized that their publications held a new purpose, that of recording the changes in species' populations and habitat.

The Lure of the Monograph

Prodigious efforts were required to finance private works by subscription in the United States. Audubon's jest about a new subscriber to the folio *Birds* being as rare as a new species gives a comic tinge to the situation, but other naturalists in a more serious tone also complained about the difficulty of attracting subscribers. The geographical distances and scattered population in America, unlike the relatively small size and dense settlement of European countries, hindered spreading the word by visual inspection or reputation. Like Wilson, Audubon and his sons were required to make repeated and extensive journeys throughout the South and New England to promote the folio works and their octavo editions. Others, such as Gray, Say, and Haldeman, relied mostly on mailed prospectuses, advertisements in magazines, and word of mouth; therefore, not surprisingly, most of their subscribers were clumped in their home region. Botanist friends attempted to sway sales to John Torrey's *Flora of North America* but admitted to the author that getting the three hundred subscribers needed for the work to break even would be difficult: "others prefer the trash Eaton is putting out." (Amos Eaton was a textbook author who remained with the Linnaean system while Torrey was championing the new "natural" system of classification.) The mention of prices would dampen enthusiasm, as even an unillustrated work such as Torrey's cost $1.50 a volume, $4.50 for the complete set, in an era when a laborer's wages averaged $1.00 a day and the popular paperback novels sold for around fifty cents. Gray's *Genera* went for $6.00 a volume, and Haldeman's *Univalve Mollusca* for $5.00 a volume. One scholar has observed that the $100 needed to buy the first octavo *Birds* (one tenth of the folio *Birds*) is the equivalent of over $1,500 in today's terms.[36]

Good-quality materials and skilled labor made these works expensive. The paper for the plates, in particular, had to be specially treated to capture ink and to be of highest quality: in the words of one author, "Nothing sets off an engraving like good paper." Authors appreciated handsome typesetting and wide margins to make a work as visually elegant as European monographs (Bachman, for example, admired Holbrook's letterpress format and recommended it for the *Quadrupeds*), but setting up and even deciphering the manuscripts full of recondite Latin phrases tested the typographer's skill.[37]

71

Creating
the Inhabitant
Image:
U.S. Natural
History
Monographs,
1825–1850

Above all, it was the labor involved in making the plates that made works so expensive. The price of letterpress printing was a tiny fraction of the costs of producing the plates, as the figures involved in production of the octavo *Birds of America* edition demonstrate. Printing 1,200 copies of the letterpress cost just over $30, while just 300 copies of one number (five plates) cost $116. Normally, the original artist was paid, as in the case of Sprague in the *Genera*, who received $6 a drawing for one plate (author-artist Audubon was an exception). Thomas Say was delighted that Lucy developed proficiency in shell drawing, as it spared him that cost. Another, often equal, expense was the lithographer or engraver who translated the design onto the printing plate; the highly skilled William E. Hitchcock, who worked on the Audubon/Bowen octavo plates, was paid $7.00 an image, while Prestele usually worked for $4.50 an image (and once modestly refused a raise Gray offered him). Hand-coloring added several cents to each plate's production. These accumulated costs of the many illustrations the monographs required (frequently over one-half the selling price) often left little profit when book agents' fees and other items were added. It took the hefty sales volume of the Audubon octavos, at 1,200-plus for the *Birds* and 2,000 for the *Quadrupeds*, for the creators to garner a significant amount of money. It is little wonder that Gray abandoned the *Genera*, hoping for projects "that made a little profit," and Haldeman, losing money, abandoned the *Univalve Mollusca*. The Binney family was reputed to have funded the 290 or so copies of *Air-Breathing Mollusks* themselves.[38]

The effort and delays involved in supervising the publications from creation to hand-coloring frustrated the authors as well. (Say once joked that he would publish a number of the *Conchology* "previous to the day of judgement.") Gray complained about Sprague's lengthy labors to fellow botanists, and Say mentioned the difficulty of finding suitable artists for continuing the *Entomology*. The process of supervising plate production at a distance—as in the cases of the Audubon family in New York and Bowen in Philadelphia, and of Gray in Cambridge and Prestele in New York state—involved much letter writing and waiting for proofs and final results. Hand-coloring was exceptionally time-consuming, as Bowen had to supervise a large team of female artists to work from corrected proofs with a demanding Audubon family ready to criticize sloppy work. Prestele and his sons often lavished so many colors and precise touches that each plate became a separate watercolor; such care meant they could color only a few plates a day. Holbrook was so eager to provide only the best possible plates that he would replace images if he later received a better specimen or if he deemed final plates beneath his standards.[39]

The stillborn and incomplete projects most hauntingly reflect the lure of publishing a monograph encompassing every U.S. species. The Haldeman, Say, and Gray works fell short of their goals of total coverage of known species, but others did not even publish their dreams. Titian Ramsay Peale, who illustrated for Say and Godman, wished for an American lepidoptera work, and George Ord planned an American zoology before Godman (Godman, in fact, probably used the same images Ord had commissioned). Perhaps most unfortunate was the failure of Charles LeSueur, the most well-rounded natural history artist working in his time, to publish his work on American fishes, as the study of ichthyology was pursued only in unillustrated catalogs, books devoted to individual states, and periodical articles until the later half of the century.[40]

The accomplishments, however, were enough to encourage naturalists. Holbrook's and Bachman's triumphant corrections of their foreign counterparts radiate their confidence that U.S. naturalists could rank with their European contemporaries. They and others more fully described new species and, more important, clarified the nomenclature of previously named species because they obtained more specimens of different ages and both sexes. Godman's observation of many individual skunk pelages ("we have never seen two alike"), for example, led him to assert that all the specimens belonged to one species and that the marking variations did not constitute separate species as European naturalists had posited.[41] The lengthy life histories provided the best stage for original observations, as the précis of habits synthesized apparently every sighting by the naturalist and his colleagues over the years. The availability of specimens again proved critical in Bachman's search to explain the opossum's gestation. American naturalists had more opportunity to press friends and colleagues to collect specimens (as in the multitude of opossums Bachman examined to find a pregnant female) and firsthand observations than the Europeans, who relied on a few estab-

lished networks and transatlantic communication. Americans used
this advantage to the fullest.

73

Creating
the Inhabitant
Image:
U.S. Natural
History
Monographs,
1825–1850

The Role of Contributors

When the subject matter of a work was known, both acquaintances and strangers would send specimens or anecdotes to the authors. W. Newcomb undertook the arduous task of trapping and
shipping Audubon a live weasel from Massachusetts, relating the tale
of how the weasel got loose in his store and became sick from eating
paint. Odgen Hammond, a New York jurist, sent another live weasel
in time for Audubon to note the change in seasonal pelage. As in the
era of *American Ornithology,* anecdotes of wildlife provided by non-
naturalists could find a place in the monographs if the observer's
veracity was assured. Thomas Kearney, who engraved the illustrations
for *American Natural History,* informed Godman that, as he was check-
ing a trap before dawn, he saw a skunk's "offensive fluid" glow phos-
phorescently ("the odor left no doubt of the animal whence it
proceeded"). Nuttall received many tidbits of bird life for his *Manual,*
such as the story of a loggerhead shrike decapitating the pet canary
of James Brown, Nuttall's own publisher and the owner of a fine ornithological library. Another friend told Nuttall that he had seen a row
of cedar waxwings pass along an insect, "each delicately declining the
offer, the morsel has proceeded backwards and forwards before it was
appropriated." Such contributions received respectful acknowledgment in the texts. Trappers and hunters could tell much from their
experiences, but Bachman warned Audubon that the tendency to exaggerate made some information doubtful: "I don't much trust the
long yarns of trappers, still they give me a good deal of information."[42]

The custom of taming wildlife for pets yielded rich evidence of
habits. Since colonial times, young birds and animals were taken from
their nests or parents to be raised unafraid of humans. Not only children but also adults enjoyed the pets. Gray and red squirrels and
birds with noted vocal abilities, such as the mockingbird and crow,
were the most popular choices, yet these histories hint that other species were invited into the domestic interior as well. The prominent
newspaper editor Gideon B. Smith of Baltimore wrote in a letter published in the *Quadrupeds* that the flying squirrel he took from an abandoned purple martin's house allowed Smith and his family to "[lay]
him on the table or on one hand, and expose the extension of his
skin, smoothed his fur, put him in our pocket or bosom, &c. he pre-
tending all the time to be asleep." Even the untamed flying squirrels

that flew into the Smith house "on a summer's evening when the windows were open" allowed themselves to be caught without biting (the family apparently harvested the squirrels one summer so the young girls could line their capes with tails the next winter). The Hartford, Connecticut, home of Daniel Wadsworth, now best known for his art collecting, held a woodchuck for more than two years. Wadsworth "began to take a particular interest in its welfare, and had a large box made for its use," in which the animal hibernated in the kitchen. After the woodchuck slept for six weeks, Wadsworth took it into the parlor and "rolled [it] over the carpet many times, but without effecting any apparent change in its lethargic condition . . . desirous to push the experiment as far as in my power, I laid it close to the fire" and saw it revive within half an hour. "An accident deprived me of my pet, for having been trodden on, it gradually became poor, refused food, and finally died extremely emaciated." An associate of Nuttall told him of a young cedar waxwing who, "abandoned by his roving parents, threw himself wholly on his protection. At large, day and night he attended the dessert of the dinnertable for his portion of fruit, and remained steadfast in his attachment to Mr. W. till killed by an accident, being unfortunately trodden under foot." The Baltimore oriole of Mrs. P. A. Messer was more fortunate in that it lived more then seven years under her protection. It would leave its cage "in very cold weather . . . fly to me, run under my cape, and place itself on my neck," and it traveled in a special cage on her railroad journeys. She claimed that the oriole "always appeared depressed and low spirited" when she was ill in bed, "often creeping under the bed-clothes to me."[43]

Messer's raising of the chick shows again the strikingly physical bridging of the species, reminiscent of the lady supplying the hummingbird sugar water from her mouth in Wilson's account: "This bird I took from the nest when very young, with three others; but, being unskilled in taking care of them, this only lived. I taught it to feed from my mouth, and it would often alight on my finger, and strike the end with its bill, until I raised it to my mouth, when it would insert its bill and open my lips, by using its upper and lower mandibles as levers, and then take out whatever I might have there for it." Messer herself contributed the observation that the peculiar action of the mandibles for gentle probing and opening made it possible for the oriole to eat peas without detaching the pod from the vine. The publisher of Brewer's edition of *American Ornithology* had presumably heard of Messer's extraordinary relationship and so requested her to furnish her ample communication, printed in a footnote to the Wilson text.[44]

Most observations and specimens, particularly in the case of the invertebrates, came from naturalists well known in other fields or persons who dedicated much time and effort to the disciplines. Several doctors, expanding upon their brief introduction to natural history in the medical school curriculum, sent Holbrook live reptiles and salamanders from across the country so that he could judge specimens from a variety of locales and establish their geographical distribution (for example, D. H. Storer sent a live and poisonous water adder from his state of Massachusetts). Academy of Natural Sciences member Jacob Gilliam collected several specimens of *Tityus* (a beetle) "in highest perfection" in Maryland so his close friend Thomas Say could supply his artists with the best model. Benjamin Say shipped his brother shells from New Jersey.[45]

Like Dr. Nathaniel Potter in the era of Alexander Wilson, the contributors often compiled notes and wrote commentary in the same scientific manner as the better-known authors. For example, Dr. John T. Sharpless of Philadelphia wrote the entire life history section of the swan in Audubon's *Ornithological Biography,* including migration, calls, and breeding periods, its flight velocity as he calculated it, and corrections of one European traveler. William Oakes of Ipswich, Massachusetts, best known for his published botanical works, had such complete records on bird sightings and migrations that he was able to inform Audubon on New England geographical distribution. A South Carolina plantation owner, J. Hamilton Couper, sent S. S. Haldeman drawings and remarks about a snail species that he had observed over time in the rice fields and captured in a jar. William Cooper of Weehawken, New Jersey, "an intelligent and close observer," had spotted a white footed mouse protecting her young while on the move by holding them to the teats by her tail—in Bachman's words, "an remarkable instance of devotion to the young." Helping a worthy project and sharing items of natural interest or curiosity motivated both the experienced naturalist and the observer who had access to wildlife for profit or pleasure.[46]

The Naturalist's Focus

General observers might adopt fledglings or notice an animal running away in the woods, but only those dedicated to pursuing natural history maintained that peculiar focus that sought after and compiled knowledge of individual species. In this era in which only a very few such as Harvard professor of botany Asa Gray and successful author Audubon made their livings from natural history, actions like extended notekeeping, collecting, and corresponding, which required long hours of study over years, not professional degrees or

75

Creating
the Inhabitant
Image:
U.S. Natural
History
Monographs,
1825–1850

positions, distinguished the naturalists from other members of society. The concept of professionalism had not yet separated the Grays, Audubons, and Holbrooks from the Sharplesses and Coupers.[47]

In contrast, non-naturalists generally experienced animal and plant life in more limited and personalized contexts. Any extended contact with fauna occurred mostly with domesticated animals as in farming and petkeeping, and so knowledge of birds or mammals would have been derived from watching a homebound bird (as in Messer's case) or the family's livestock. The cultivation of crops and flowers necessitated knowledge of life cycles, although for a limited number of species. Only such special cases as trapping and hunting required intimate knowledge of animal habits.[48]

The naturalists desired more information about all the species in their particular field of interest and therefore frequently lived with specimens in their home for the specific purpose of scientific observation. Often they were treated as pets, as in the cases of baby raccoons romping with Godman's children in his parlor ("without doing any injury even to the youngest") and a Canada porcupine who ate sweet potatoes from John Bachman's hand. The Bachman household held, among other creatures, the many opossums, "generations of flying squirrels," and two pugnacious white herons who ran through a cat with their beaks (finally, their menacing attitude toward the Bachman children caused them to be killed, with their stuffed carcasses going to museums). Haldeman and Binney likewise studied snails in jars at their homes, and Holbrook kept many frogs, snakes, and turtles. His tortoises lived "for many years" in the lower part of his garden and enjoyed frequent sunbathing. He held one "fairly tame" tree frog in a box filled with earth in his study; when the weather was warm, "it enjoyed rain baths" from Holbrook dripping water onto the skin, "nor did he fail during the extreme heat of the summer to repair to it frequently." The naturalists' duty to study habits was, however, paramount in their relationship to these household inhabitants. Keeping live specimens enabled Audubon to mark the weasel's change of coat and Holbrook to note the food and the sounds of the lesser-known amphibians. Bachman saw the opossum mother guide the tiny young to her pouch as he worked on other matters in his study.[49]

Outdoor observation provided the bulk of evidence. Constant hunting and quiet watching allowed the naturalist eventually to witness most of the life cycle of even the shiest species, including the brief moments of coition such as those in Godman's raccoon accounts. Audubon tactfully but definitively conveys his sighting of the Virginian rail's mating one morning:

The notes of the Rail came loudly on my ear, and on moving towards the spot whence they proceeded, I observed the bird exhibiting the full ardour of his passion. Now with open wing raised over its body, it ran around its beloved, opening and flirting its tail with singular speed. Each time it passed before her, it would pause for a moment, raise itself to the full stretch of its body and legs, and bow to her with all the grace of a well-bred suitor of our own species. The female also bowed in recognition, and at last, as the male came nearer and nearer in his circuits, yielded to his wishes, on which the pair flew off in the manner of house pigeons, sailing and balancing their bodies on open wings.[50]

77

Creating
the Inhabitant
Image:
U.S. Natural
History
Monographs,
1825–1850

This observation at close range inspired detailed characterizations with attributes and actions often shared with humans. Holbrook noted the *Coluber Alleganiensis* "in confinement seemed of an exceedingly mild and gentle disposition; forming in this respect quite a contrast with its fellow prisoners, two individuals of the common Black Snake (*Coluber constrictor*), who maintained at all times their original wildness." Holbrook thus used empirical evidence and comparison to back his humanlike distinction between wildness and gentleness. Bachman claimed the porcupine would beg for a treat "with a mild and wistful look," again attaching human traits based on his daily experience. Ornithologists, in particular, went so far as to promote their subjects as exemplars. As in Audubon's hummingbird passage, Nuttall extolled the virtues of birds as loving parents and devoted spouses (claiming "Conjugal fidelity and parental affection are among the most conspicuous traits of the feathered tribes"), and both naturalists often explicitly give them as models for human conduct. Nuttall echoes Wilson's dislike of ignorance and superstition in his discussion of the barn owl's cry: "They are not left by nature as spectacles of derision, but have their calls of complaisance, of recognition, and attachment, which, though discordant to human ears, are yet only ordinary expressions of agreements and necessity."[51]

The naturalists did not commend every trait. In the attempt to portray the species truthfully, they included unpleasant events like intraspecies warfare, as when Nuttall repeated Wilson's observation on the battling titmice. Nuttall also admits that the "constancy" of most mating birds lasts only the season, unlike the birds of prey, who usually pair for life. Audubon notes, after the "pleasing sight" of redpoll linnets passing seeds to one another, this jarring observation: "Occasionally, however, they shewed considerable pugnacity, and one would drive off its companion, inflicting some smart blows upon it with its bill, and uttering a low querulous chatter."[52]

These naturalists did not envision the close integration of a species' many members with their habitat and with other species' populations as today's biologists do, but literally focused on the single specimen with background blurring, no matter how small the object of attention or how fatigued the observer. One anecdote from Thomas Say tellingly reveals this fascination with a single specimen representing the species. After a particularly difficult few days during the Long Expedition, the Konza Indians offered shelter to his band of "hungry and fatigued" travelers. "Whilst sitting in the large earth-covered dwelling of the principal chief, in presence of several hundred of his people, assembled to view the arms, equipments, and appearance of the party, I enjoyed the additional gratification to see an individual of this fine species of *Blaps* [dung beetle] running towards us from the feet of the crowd. The act of empaling this unlucky fugitive at once conferred upon me the respectful and mystic title of 'medicine man,' from the superstitious faith of that simple people."[53]

Naturalists weighed and evaluated physical appearance within and among species as a matter of course. This tendency to judge individuals in terms of a standard physical appearance became comical in one example of Native American and explorer contact. Thomas Nuttall accompanied the 1810–1811 expedition to the mouth of the Columbia River that later became the basis of Washington Irving's *Astoria*. A companion told fellow traveler Henry M. Brackenridge this incident, which Brackenridge later published in his memoirs: "The party [confronted by Indians] was on the point of firing; while every one was in momentary expectation that this would take place, Nuttall, who appeared to have been examining them very attentively, turned to Miller, 'sir,' said he, 'don't you think these Indians much fatter, and more robust than those of yesterday?' " Nuttall's other actions gained notice in Brackenridge's journal and *Astoria* as examples of the naturalist's concentration on his duties to the detriment of the expedition. His habit of darting from the boat to gather plants caused delay, and it was found in one instance of danger that his gun was stuffed with dirt, as he had used it to dig up plants. This intrepid and important naturalist unluckily became a prototype for the other-worldly naturalist, Obed Battius, in James Fenimore Cooper's *The Prairie* (1833). This comic character, who during an arrow attack slips from his place of safety to seize an unfamiliar plant, automatically mutters the order, class, and genus of every living creature he encounters ("Class mammalia, much less a man" as he fails to recognize an Indian warrior).[54]

Creating
the Inhabitant
Image:
U.S. Natural
History
Monographs,
1825–1850

The financial success and pleasures of non-naturalists were often denied to the naturalists, but their constant watchfulness over time and place brought forth special pleasures. Others may have enjoyed the outdoors, but the naturalist's sustained presence converted that enjoyment into a more lasting enjoyment. In the words of Thomas Nuttall, who gained financial independence only through an inheritance toward the end of his career, "privations to him are cheaply purchased if he may roam over the wild domain of primeval nature . . . for thousands of miles my chief converse has been in the wilderness with the spontaneous productions of nature; and the study of these objects and their contemplation has been to me a source of constant delight." Non-naturalists experienced empathy and interest in interacting with pets, but the naturalist's more frequent encounters with members of a species outside the home and in unfamiliar territories reinforced the tendency to meld individuals into one entity, often deemed a "friend." Frequently Audubon and Nuttall cite the song of birds as heartening them during difficult journeys and perplexing times. Audubon noted the reassurance that the cardinal's customary late-afternoon songburst gave him: "How pleasing is it, when, by a clouded sky, the woods are rendered so dark, that were it not for an occasional glimpse of clearer light falling between the trees, you might imagine night at hand, while you are yet far distant from your home—how pleasing to have your ear suddenly saluted by the well known notes of this favourite bird, assuring you of peace around, and of the full hour that still remains for you to pursue your walk in security!" Nuttall's biographers felt that his truest friends in the New World were the plants and birds he knew so well: "Wherever he went, whether in the valley or on the mountain, by the shores of the sea or the margin of the quiet stream, he felt surrounded by old acquaintances, his dearest flowers."[55]

One reviewer expressed the quality of revelation that the naturalist's activities brought: "The common observer of nature sees a bird or an insect, admires its shape and colour, but takes no further notice of it. Not so the naturalist; he must follow it to its haunts, mark its movement, and learn all its peculiarities. By constant attention, he sees these, when to ordinary beholders they are invisible." The revelations include Godman's elucidation of the shrew mole's structure, Bachman's new understanding of the opossum's nursing of its young, and Holbrook's insight into the helpfulness of snakes in reducing the pest population. To Bachman and William Cooper, the white footed mouse so ingeniously sheltering her young seemed miraculous. Naturalists of this era freely linked their insights to God's craftsmanship, and, as Wilson had praised the work of the barred owl's feathers, expressed their wonder and sense of privilege in witnessing the perfection.

Their surrounding culture also extolled Nature as a pathway to God alongside more formal religions. The belief in America as "Nature's Nation" fostered a quasi-religion expressed in literature and art where its adherents "found" the Godhead in His physical manifestations.[56] Other Americans adhered to the "belief" through painting or poetry or in a generalized appreciation of the surrounding beauty. Naturalists, through their studies, sought and found in Nature the functions associated with spiritual fulfillment. Their rewards were a sense of purpose, an awareness of revelation, and a feeling of attachment (as in the plants who were Nuttall's "dearest friends" and the cardinal cheering Audubon).

Language and Imagery: The Distinctive Discourses

The naturalists' language clearly separates their speech from that of other pioneers. Latinate nomenclature was the standard for universal usage (in *The Prairie*, Battius's real name is Bat, but he prefers it Latinized), while the common or "local" names that confused Wilson continued to reign in everyday life. Fish, for example, were particularly confusing to those trying to write national works, as one name like *sheephead* was frequently used for several fish along the Atlantic coast. In the state of New York alone, the copperhead snake was also known as red adder, dumb rattlesnake, copper-belly, red viper, deaf adder, and chunk-head. The author of the zoological portion of the New York state survey, James DeKay, hoped that the English names he proposed in addition to the Latinate ones would become the standard, because the scientific community depended on information from non-naturalists: "the greater part of our knowledge of the habits of animals is derived from persons unskilled in natural history." Proposals like these and the similar efforts of English naturalists to standardize common names earned the ire of S. S. Haldeman, who in his pamphlet *On the Impropriety of Using Vulgar Names in Zoology* dismisses the approach as futile: "vulgar names being of no authority, no necessity exists to change them." He contended that the varied local usage was too entrenched (his knowledge of the Pennsylvania German names in his home district of Lancaster reveals his appreciation of the diversity of expression that non–English speaking groups contributed). He urged scientists to focus efforts on standard Latin nomenclature, a difficult enough objective because of variety within species, new discoveries, and new classification systems. The use of common

names on plates (as in the octavo *Birds*) particularly irritated him.
Both Haldeman's and DeKay's positions ironically were confirmed in
the succeeding generation. Scientists did concentrate on standardiz-
ing nomenclature rules, most notably through the efforts of the
American Association for the Advancement of Science in the 1870s.
However, most scientific works gave one or two of the most frequent
"common names," thereby accelerating the standardization of English names and the decline of the other "vulgarities" like chunk-
head.[57]

81

Creating
the Inhabitant
Image:
U.S. Natural
History
Monographs,
1825–1850

The natural history artwork genre shared the inherent focus and
conventions that differentiated it from other representations of flora
and fauna. Botanical art shows a wider range of plants than the popular and closely related genre of flower painting, in which fashionable
flora like roses, pansies, and lilies-of-the-valley dominate. Botanical
illustration also displays the unaesthetic details, such as gnarled roots
and dissected parts. Representative parts of the anatomy are fully displayed in zoological illustration, whereas in animal painting the frequently active figures are contorted without such restraints. Again,
domestic animals like sheep or the most popular wild animals like
the lion, not hitherto undiscovered species, were the primary subject
matter (Illustration 2.7).

The adherence to this distinctive natural history "visual discourse" unified the monographs' participants from varying backgrounds. Illustrators trained in various fields made natural history
drawings. For example, Benjamin F. Nutting, who specialized in invertebrates, had been a landscape artist before working for Augustus A.
Gould; William W. Wood, who illustrated Say's *Entomology*, probably
painted miniatures. Nutting's quickness in learning the genre
through copying plates pleased Gould: "Nutting has copied the figures of Valenciennes [a French invertebrate specialist] in such a manner that you would think he must have cut them out of the book."
Artists employed by commercial lithographer Peter Duval drew for
Holbrook. Their established manual dexterity and ability to copy realistic forms especially allowed the transition to this genre. For example, Albert Newsam, who drew several turtles, was famed for his ability
to draw detail, his concentration aided by his deafness; a contemporary wrote, "Newsam's principal talent was his copying. . . . It was with
difficulty he could avoid copying minutely."[58]

As the genre relied on mimicking, in relatively flat images, objects set directly in front of the artist, the lack of historical knowledge,
perspective, and anatomy (in other words, the background necessary
for landscape, portrait, and history painting, taught in art schools or

in workshops and learned over years) was not a particular drawback. Moreover, no schools or manuals geared to the genre existed, so there were no professional barriers to discourage women and working naturalists from learning the genre. The invertebrate specialist Haldeman in a few cases and, of course, Audubon illustrated their own works. Maria Martin sketched monuments for her niece's scrapbooks before beginning to copy butterflies from Say's *Entomology* in preparation for drawing plants and insects to serve as background for Audubon's *Birds*. The artists achieved competence at different times in their lives. Titian R. Peale and the Lawson offspring breathed early the familial natural history atmosphere; Nutting, Martin, and other artists learned as adults. The Reverend John Henry Hopkins, bishop of Vermont, who gave Holbrook his salamander drawing, cut his illustration teeth as a youth coloring Wilson's *American Ornithology* plates; he later published his own drawing manual of Vermont scenery.[59]

Illus. 2.7. "Exercise No. 206 and 207." Wood engraving by anonymous artist for *Drawing for Young Children* (New York: Charles S. Francis, 1853). (Courtesy of Winterthur Library, Printed Book and Periodical Collection.)

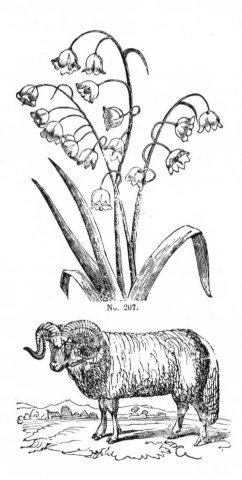

No. 207.

Adherence not only to natural history's general conventions but also to those specific to the era unified the contributions of multiple artists without noticeable discrepancies. Indeed, the Audubons and Gray had authorial control over their illustrations. However, the many artists for *North American Herpetology* conveyed both anatomical accuracy and the sense of life or "spiritedness," all the more surprising considering that Holbrook could not supervise personally the geographically scattered individuals on a regular basis. The various artists for Godman and Say, working at different times and locations, likewise produced drawings that shared similar compositions. They may have known previous works in the projects, but more important, they understood the special requirements of focusing on the specimens and employing background as a foil.

Those who translated the drawings onto plates similarly geared their techniques to suit the genre's requirements. Alexander Lawson engraved scenes for general book illustrations in a much looser fashion than for his natural history mode, and John T. Bowen lithographed landscape prints and advertising literature and completed in 1844 the portraits that constituted the magnificent folio edition of *History of the Indian Tribes of North America*. Both artists became proudest of their skill in producing natural history illustrations. In a famous conversation recorded by art historian William Dunlap, Lawson dismisses Audubon's drawings as faulty because of the position of the feet, among other inaccuracies: "This leg does not join the body as in nature. . . . we in Philadelphia are used to seeing very correct drawing." Lawson thus believed that his experience in engraving for Wilson and Bonaparte had taught him avian anatomy. Bowen once told the Audubon family that he would engrave a *Quadrupeds* design on stone for the fourth time because he, like them, was not satisfied with its accuracy. The Bowen firm, in fact, eventually specialized in the Audubon octavo editions and other high-quality natural history books.[60]

The European Comparison

Despite limited distribution in Europe due to the difficulties in promoting and shipping the works, the American monographs generally received favorable reviews in foreign scientific journals. More importantly, European naturalists conferred approval on the species first described by such Americans as Thomas Say and Alexander Wilson by acknowledging them in their own publications. Prideaux John Selby in the text for his *Illustrations of British Ornithology* (1818–1834) often compares his specimens' physical characteristics with those of

American analogues given by Wilson to determine their familial relationship. William Swainson, an eminent and prolific English naturalist, violently and publicly disagreed with American conchologist Isaac Lea's classification in the division of *Unio* (freshwater shells), for example, but acknowledged that Lea's descriptions of one thousand new species "deserve the thanks of every naturalist."[61]

The innovations of life histories and the living bird in particular attracted the most comment in European scientific circles. The ongoing requirement to name and place in a coherent order the thousands of specimens pouring into the foreign natural history collections from Asia and the Americas ensured that nomenclature and taxonomy remained primary concerns in the late eighteenth and most of the nineteenth centuries. As historian Paul Farber has noted, however, some European naturalists were interested in local fauna and their habits in disciplines otherwise dominated by these classification concerns and therefore appreciated the field studies of Wilson and Audubon. John Wilson, himself an ornithologist and the editor of *Blackwood's Magazine,* likened Alexander Wilson's and Audubon's absorption in the "living, breathing" bird to that of then-popular English authors James Rennie and Robert Mudie, who also specialized in observing commonplace species. John Wilson proclaimed that accounts of habits and anecdotes could more effectively popularize the subject and contribute to ornithology than the dry discussions of classification and the endless catalogs listing only brief physical descriptions favored by other contemporary ornithologists: "Proficients, no longer confining themselves to mere nomenclature, enrich their work with anecdotes and traits of character . . . imbu[ing] bird biographies with the double charm of reality and romance." His British editors, Jameson and Jardine, so admired Alexander Wilson's prose that they reprinted every word; Prideaux John Selby praised his graphic style and similarly included many firsthand experiences of tamed and wild birds.[62]

Audubon's influence lasted well beyond his sensational debut in 1826 as an interesting "American Woodsman" delighting acquaintances with his imitations of owl calls. Selby and Sir William Jardine took lessons in drawing birds from him, and Selby's subsequent plates in *British Ornithology* show increased interest in varied poses. Swainson, also an accomplished artist, is believed to have shown greater animation and variety in his bird compositions owing to Audubon's influence. MacGillivray, his *Ornithological Biography* editor, drew magnificently designed compositions in the style of Audubon that were not published. The publicity generated by Audubon's *Birds* revitalized the folio ornithological tradition. John Gould and Edward Lear, both extremely familiar with *Birds* and Audubon himself, produced imperial folio monographs in the early 1830s. Gould, labeled in his lifetime "the British Audubon," continued to produce hand-colored folio books until his death in 1881. John Gould's multiple-figured compositions, often showing families against a suggested habitat, drew their inspiration from Audubon. In fact, Audubon's friend, the popular author Charles Wilkins Webber, accused Gould of copying Audubon's figures. This allegation, that Gould traced and slightly altered positions in his *Birds of Europe* from figures in *Birds of America*, whether accurate or not, itself reveals the transatlantic interest in lifelike postures different from the traditional side-view stance[63]

85

Creating
the Inhabitant
Image:
U.S. Natural
History
Monographs,
1825–1850

Viewed in retrospect, the lack of theoretical novelty compared to European study enhanced the Americans' long-lasting contributions. British naturalists such as Swainson and MacGillivray were expending great effort on their own systems of nomenclature: Swainson developed the odd quinary system in the almost mystical belief that classes, orders, and genera occurred in units of five; MacGillivray based his system on the emphatically physical evidence of gastrointestinal tracts. Both systems won few adherents and created only more confusion for zoologists and their readers. Nuttall, Godman, and Say in various fashions followed basic European precedents but did not make the creation of new classification methods their métier. In Godman's view, a system could not capture nature; Nuttall's work in naming many new plant species similarly made him appreciate infinite variety and disavow one perfect classification. Detailed physical descriptions accompanied by accurate illustrations like those in the publications of Say, Audubon and Bachman, and Wilson survived despite various reconfiguations of nomenclature because their joint visual and textual precision enabled zoologists to understand the species. Jardine, for example, could append newer systems to Wilson's descriptions more than two decades later because Wilson still described mostly valid species: "If some plates and illustrations may vie with it in finer workmanship or pictorial splendor, few indeed can rival it in fidelity and truth of delineation."[64]

Destruction and Disappearance

Because of the empathy nourished by continual contact, the death of their subjects sometimes distressed the naturalists. Nuttall called neighbors to saw down a bough in the aid of a female Baltimore oriole who became entangled in her nest. After praising the song of the red-eyed vireo that is effortlessly poured out while seeking insects "as a messenger of harmony to man *alone*," Nuttall interjects a warning to his readers: "Wantonly to destroy these delightful aids to sentimental happiness ought therefore to be viewed, not only as an act of barbarity, but almost as a sacrilege!" Holbrook likewise deplored the "universal prejudice against serpents" because it encouraged persons to destroy creatures he thought fascinating and often helpful to humankind. These authors, not influenced by the modern desire for scientific objectivity, freely expressed their opinions through the ample format of the life histories.[65]

This admiration clashed with the necessity of killing birds for research purposes, a tension that often reverberates in the texts. Nuttall so appreciated his subjects that he reportedly refused to shoot them; Audubon, on the other hand, as modern critics relish revealing, was an expert shot, who coolly mentions his harvests in the *Ornithological Biography*. The rapidly blossoming sentimentalism in transatlantic culture that influenced many of Audubon's descriptions made such specimen-gathering suspect. It seemed increasingly to his readers and to Audubon himself that he would be accused of killing beloved kindred beings rather than dispensable objects. Passages reveal his awkward straddling of the roles of working naturalist and empathic admirer, as in his praise of a Mississippi kite's courage in retrieving her young after he had shot at her: "I wished I had not discovered the poor bird; for who could have witnessed, without emotion, so striking an example of that affection which none but a mother can feel; so daring an act, performed in the midst of smoke, in the presence of a dreaded and dangerous enemy. I followed, however, and brought both to the ground at one shot, so keen is the desire of possession!" Future ornithologists would struggle similarly to justify their hunting in the face of public appreciation of songbirds, especially as stricter enforcement of bird preservation was enacted around the turn of the century.[66]

Some commentators have posited that pre-Darwinian images and texts in general embody a statically atemporal system because no thoroughgoing theory of evolution had overtaken the traditional position of the once and final creation by God. However, life histories allowed naturalists to record their observations of the dwindling of species numbers due to, in their words, "encroachment of civilization." Audubon's famous passage on the changes in the Ohio Valley in the first quarter of the century—"That these places deserve a Washington Irving or Cooper to commemorate them ere they leave"—perhaps has given him the unwarranted position of being one of the few voices of concern. As with the Native Americans' "disappearance," many other commentators saw less fauna over years and linked the decrease to hunting and changes in habitat. The requirement of defining the geographical distribution of each species particularly encouraged authors to consider these issues. Audubon and Godman discussed overhunting of game mammals; Godman comments, for example, that the cougar "appears to diminish in direct proportion to the advances of the human race, and the extension of cultivation. . . . Their race even finally extinguished."[67]

As settlement extended into the West, local authors saw the same patterns of diminution repeat. Audubon, who lived in Kentucky in the early part of the century, admits that by 1847, "no [elk] are to be found within hundreds of miles of our then residence." William Case of Cleveland, Ohio, wrote in his contribution to the Canada porcupine description published in the *Quadrupeds* that "no more than ten years ago one person killed seven or eight in the course of an afternoon's hunt for squirrels, within three or four miles of the city, while now probably one could not be found in a month." A prejudice, "the extreme hatred all hunters bear them on account of the injuries their quills inflict on their dogs," had made them too frequent targets— "They are rapidly becoming extinct."[68]

The theme of loss had become so widespread that even those who did not live in the United States enunciated it. The suddenness of change from absolute wilderness to cultivation struck with peculiar force the European naturalists accustomed to a more humanized landscape and the absence of a "native" population. Sir William Jardine, for example, reflected in his 1832 preface to *Wilson's Ornithology* that Thomas Say a decade earlier had noted changes to the Indian population through encounters with whites. Jardine so appreciated the ongoing transformation that he praised Wilson's work for its record of the past before "The culture of the country destroy[ed] the natural productions, and introduc[ed] others not adapted for them. . . . the great and indiscriminate destruction of the different species whether for food or as different articles of commerce."[69]

For the *Quadrupeds,* the Audubon family in the 1840s melded zoological illustration with contemporary symbols of landscape destruction to create iconography suggesting the changes in habitat. Many plates feature the unspoiled wilderness of mountains and prairies, but some include peacefully pastoral barns and farms at a distance. As tellingly, the tree stump, indicating white settlers' uprooting of nature, and traces of Indian settlement such as tents and fires (and, in a few cases, Indians themselves) represent the changes that had occurred and would continue. Indians symbolize the vanishing past, as the many contemporary references demonstrate. The numerous stumps in the "Say's Squirrel" plate with the cabin in the distance communicate the loss of the forest and the newer habitat that had come (see Illustration 2.2). Victor Gifford Audubon, the landscape specialist, would have known these visual metaphors used by his fellow East Coast landscape artists to indicate the loss of wilderness; he deliberately chose them to link visually species and habitat diminishment.[70] The inclusion of species disappearance would only increase in natural history literature as transformations of habitat accelerated in the decades to come.

Naturalists thus took upon themselves to publish their own favorite sections of the plant and animal world in the most painstaking method possible. Delicate hand-coloring, precise engraving and lithography, and hours of labor were deemed essential to making works worthy of the subjects, groundbreaking in the disciplines, and competitive with those of Europeans. Because of the vagaries of subscription publishing, more than a few projects failed to reach completion. The more successful productions—Audubon's *Birds,* Godman's *American Natural History,* Bachman and Audubon's *Quadrupeds,* Holbrook's *Herpetology,* and Gray's *Genera*—did expand the American illustrated monograph tradition initiated by Wilson and included the other disciplines.

The life history format contained the wealth of observations made by the naturalists themselves and their helpers. The "inhabitant" advantage is most evident in the number of specimens encountered and habits revealed, the range of which is captured with evident pride in the textual descriptions. The richness of species characterizations—Holbrook's "bold" watersnakes, Nuttall's "bleating" hummingbirds, and Say's moths "reluctant in flight," for example—also belies the years of observation at close range. The illustrations extend the "bird living in nature" iconography suggested by Wilson to plants, as in the *Genera*'s springing flowers, and to mammals, as in Audubonian squirrels against a background. The abundance, vitality, and variety of shell species are expressed through the conchological designs. Illustrators like Isaac Sprague and engravers and lithographers like Joseph Prestele, the Lawson family, and John T. Bowen became thoroughly skilled in the naturalist's imagery as they themselves crafted the illustrated plates. The European appreciation of Americans' lifelike iconography and concentration on habits enhances the Americans' contributions.

The creation of the monographs in large part depended on non-naturalists for very material support. Subscribers financed the continuation of the projects, and friendly non-naturalists contributed information. In the following generation of natural history monographs, the mechanics and therefore expenses of publishing remained unchanged, and naturalists would continue to rely on correspondents. A new source of financial support—government—would play a significant role, however, and the westward emigration across new territories would yield valuable local observers.

89

Creating
the Inhabitant
Image:
U.S. Natural
History
Monographs,
1825–1850

3

Nature Writ Large and Small:
Local, State, and National Natural Histories,
1850–1875

What a pest, plague & nuisance are your official, semi-official & unofficial Railroad reports, surveys &c. &c. &c. Your valuable researches are scattered beyond the power of anyone but yourself finding them. Who on earth is to keep in their heads or quote such a medley of books—double-paged, double titled & half finished as your Govt. vomits periodically into the great ocean of Scientific bibliography.

William Hooker, Letter to Asa Gray, 26 March 1861[1]

These local catalogues are not as highly appreciated as they should be. The day will come when they will all be collected and bound in a volume labeled the "Ornithological Bible" which will be handed down with pious care to all coming generations.

Harmon A. Atkins, Letter to Edgar A. Mearns, 15 December 1879[2]

*W*eighty, amply illustrated tomes describing seemingly every specimen collected in the great government expeditions of national expansion, together with more modest publications listing the presence of flora and fauna in localities, characterize natural history publications before and after the Civil War. The tremendous increase in national territory due to annexation of western lands previously held by Mexico and Great Britain stimulated surveys to map and describe for the nation its new parts. These were conducted first by the Army Corps of Topographical Engineers and then by its successors in government exploration, the Army Corps of Engineers and the United States Geographical Surveys. The economic potential for mineral and agricultural wealth as well as curiosity about the unknown territory's possibilities for becoming "home" to U.S. citizens motivated the public support. Nonetheless, botany and zoology benefited immeasurably as the goal of producing the best-quality government surveys possible ensured the relatively complete and expensively illustrated works that hitherto had proven elusive. Although these works were government publications, private individuals—naturalists as well as artists and printers—molded them, for identifiable members of expeditions collected discrete specimens and noted their characteristics, with the leading naturalists and artists describing the same specimens for final publication. Thus, public purpose underwrote private effort.

91

Nature
Writ Large
and Small:
Local, State,
and National
Natural
Histories,
1850–1875

Private publishing efforts also continued in the pattern of the ambitious national monographs, based on the research of individuals who studied intensely the natural world around them. As more persons of European descent permanently settled throughout the United States, more inhabitants contributed information from diverse locations and, stimulated by their new environs, themselves published local lists of specialized plants or animals. The growing information on geographical distribution, supplemented by nationwide contributions of specimens to the new museums (in particular to the Smithsonian Institution under the leadership of Spencer Fullerton Baird), stimulated American naturalists' contributions to that crucial issue of nineteenth-century natural history, the origin of species. American naturalists previously had been interested in geographical distributions, but mostly in order to distinguish U.S. from European species and to give a clearer indication of species' habitat. Describing various specimens from the vastly varying regions now encompassed by national boundaries gave rise to specific instances of how different species originated from common ancestors and how they were still evolving. The American landscape and its products were realizing the promise of distinctive intellectual riches.

Local Natural Histories

Naturalists like Benjamin Smith Barton and Jacob Bigelow had been producing local records of plants growing near Philadelphia and Boston, respectively, since the early republic. Such lists concentrated on one feature of the living world in a locality, such as plants, shells, or birds. The observer listed all species he or she noted, sometimes referring to others' findings. Compiling such recordings was deemed as important as preserving human records. In Barton's words, "They form, next to the living, the best picture of the country . . . [and are] the most precious monuments of natural history that can be bequeathed to an inquisitive and enlightened people," as they suggested the relationship between climate, plant growth, and seasonal time (the study known as phenology) as well as signifying to future generations what creation had existed. The making of the lists provided personal satisfactions also, particularly for those new to an area. Noting the yearly cycles of bloom and fruition and the annual migration of birds attached the observers to the environs.[3]

The Observers

Although the major botanists and ornithologists had published many species by midcentury, these local observers believed in the enduring importance of their work. The young Spencer Fullerton Baird, on the threshold of his career as secretary of the Smithsonian Institution and the most influential naturalist in the latter half of the nineteenth century, wrote in his 1844 "List of Birds Found in Cumberland County, Pennsylvania" that "many blanks remain to be filled up; much still depends upon local observation, and many acts must be gathered by observers of small districts." He claimed that the ornithological giants, Audubon and Wilson, had missed species' nesting habits and attachment to localities because they were always traveling. "[M]en who have the objects of their attention and inquiry constantly before their eyes" in one place, he felt, may better judge such aspects. Indeed, these observers studied their areas for years at a time; Baird and his brother William compiled their list over a four-year period, while others kept detailed notes for decades.[4]

The expansion of the United States meant that more inhabitants observed in greater geographical diversity. For example, a Swedish immigrant, Thure Kumlien, settled in Koshkonong, Wisconsin, in 1844 with years of experience in Scandinavian botany and ornithology, but not until he purchased Thomas Brewer's edition of *Wilson's Ornithology* in 1849 did he know the names of the new birds around him. Armed with this fresh knowledge, he wrote his first list of birds and their spring arrivals in his area. In a wonderful completion of the naturalists' circle, the Bostonian Brewer (who was also a newspaper editor) read an article reprinted from a local Wisconsin paper that praised Kumlien's list and bird collection and began a correspondence with him. Kumlien himself contributed to a nationally distrib-

uted magazine, *Field and Forest*, his observations on the breeding of
Wilson's phalarope. That the male of Wilson's phalarope incubated
the eggs (the male's belly was naked, as was usual for nesting birds)
and that the females "often two at a time" pursued the male during
mating season prompted the rising ornithologist J. A. Allen to marvel
at these "quite novel facts . . . unique in the history of our birds."[5]
Dr. Harmon A. Atkins of Lock, in Ingham County, Michigan, began
a correspondence with Edgar Alexander Mearns, a young doctor in
New York who later served on national expeditions, after hearing of
Mearns's local list, *A List of the Birds of the Hudson Highlands*, in 1879.
Atkins's replies to Mearns's queries reveal his years of recording
spring migrations: "It gives me much pleasure to be able to present
to you the dates of the appearance of the Golden-crowned Thrush
for the past 23 years. It has cost me a good deal of labor to go over
my notes and get the exact dates but the disposition to impart knowledge to a fellow laborer in the same field, makes the work light."[6]

93

Nature
Writ Large
and Small:
Local, State,
and National
Natural
Histories,
1850–1875

A future minister born and raised in Massachusetts, Moses Ashley
Curtis (1808–1872), had little knowledge of southern botany on his
arrival at Wilmington, North Carolina. Looking for a new hobby, he
sent specimens to William Darlington, a Philadelphia botanist, and
began field study in 1832. Within three years, he wrote a fifty-nine
page "Enumeration of Plants Growing Spontaneously around Wilmington, North Carolina," with 1,031 species—a number not surprising to him, as "no section of the Union, of equal extent, contains
such a rich and extensive variety of plants as are to be found about
Wilmington." Decades of extended study while he was a clergyman
in the Carolinas led to two later catalogs listing thousands of species
published in conjunction with state surveys. The inscription on his
memorial tablet bespeaks the pride he and his family felt in this
achievement: "A world authority on fungi: In 1867 his published list
of 4,800 North Carolina Plants was the largest North American re-
gional list."[7]

Nor were other segments of biota forgotten. The young Edward
S. Morse, on the threshold of his varied career as invertebrate specialist, artist, and Far Eastern porcelain collector, by age thirteen had
studied and collected land snails, in part inspired by John W. Chicker-
ing's published list of his collection, *List of Marine, Freshwater and Land
Shells Found in the Immediate Vicinity of Portland, Maine* (1854). In 1864
Morse published his first book on the land snails "known to inhabit
the state," *Observations on the Terrestrial Pulmonifera of Maine, Including
a Catalogue of All the Species of Terrestrial and Fluviatile Mollusca Known
to Inhabit the State*, illustrated by himself.[8]

Such local lists were usually confined to brief descriptions, but a few authors developed fuller life history profiles based on their observations. Thomas Gentry, in his *Life Histories of Birds of Eastern Pennsylvania* (1876), described in unprecedented detail the nesting structures and habits "hitherto not published," based on his study of over five hundred nests and "many pairs of the species whose habits . . . have been closely and diligently watched" over six years. The intense focus enabled him to go beyond discussion of the typical nests to speculate on the probable causes for "markedly deviant" nests. Like Kumlien, he commented on gender roles in nest building and incubation, or in his words, "the labor of nidification whether performed by the male and female separately, or conjointly" and the "periods of incubation . . . whether accomplished by the female exclusively, or with the assistance of the male." The unusual ability of the Venus's flytrap and pitcher plant to capture insects had long fascinated naturalists, who wondered whether the insects served as subsistence. The botanist Curtis stated his belief in his Wilmington enumeration that the pitcher plant did not crush the flies as supposed (because "I have often liberated captive flies and spiders, which sped away as fast as fear or joy could hasten them") but rather dissolved them in its "mucilaginous" substance, thus opening the "possibility of their being made subservient to the nourishment of the plant." Hours of lying on his stomach in swamps, waiting for the plant to capture insects, produced his insights.[9]

The Publications

Financial constraints determined that most local histories were unillustrated (unless the author was an artist, as in the case of Morse, who drew and engraved on wood for his publication). One unusual and visually magnificent case of local fauna publications, however, featured illustrations as its raison d'être. The Jones family of Pickaway County, Ohio, collaborated on text and plates for *Illustrations of the Nests and Eggs of Birds of Ohio*, issued between 1879 and 1886. Although it technically falls outside our time period, its admirers noted that it followed in the great hand-colored folio tradition of Audubon. Genevieve Estelle Jones and her friend Eliza J. Schulze conceived the idea of not only sketching but drawing on lithographic stone actual nest and eggs specimens collected by Jones's father, Nelson, and brother, Howard, despite neither woman's having had previous experience in lithography. Presumably their mutual love of bird life animated the family's ambition. Nelson, an avid game hunter, studied Wilson, Audubon, and more recent ornithological literature; Genevieve and Howard shared a lifelong interest in bird behavior, beginning in childhood with raising a variety of wild birds in cages. Howard, who wrote the accompanying text, and his father sought and harvested nests for illustration during the comparatively brief spring-time window of opportunity in the midst of their medical practice. Howard Jones claimed to have studied so many nests of a single species that he could tell from the structure the individual maker's experience, "This is the nest of an old bird."[10]

Schulze withdrew from the project in 1880. Genevieve died shortly after the first part was completed, but her mother, Virginia Smith Jones (1827–1906), learned to draw on stone to finish the book dedicated to her daughter's memory. Virginia Jones and three other female artists colored the ninety copies of fifty-eight life-size plates depicting the individual nests Howard had harvested. Howard emphasizes that these specimens were drawn "fresh, rather than be taken from his cabinet," and indeed the robin plate shows a branch and the weaving green tendrils surrounding the nest with eggs still in situ (Illustration 3.1). The image of the Traill's flycatcher's nest in-

95

Nature
Writ Large
and Small:
Local, State,
and National
Natural
Histories,
1850–1875

Illus. 3.1. *"Turdus migratorius.* Robin." Lithograph by Virginia Smith Jones, colored by Josephine Klippart, for Jones, *Illustrations of the Nests and Eggs*, vol. 1, Plate 8. (Courtesy of the George Peabody Library of Johns Hopkins University.)

cludes the still legible piece of newspaper incorporated into the structure, again emphasizing the individuality of the nest. The study of eggs and nests may have lost its prominence in the natural sciences with the rise of molecular biology and genetics, but, perhaps on some level influenced by the prevailing cultural interest in new formulations of "domesticity," it flourished in the nineteenth century because of such individuals' fascination. The plates' renditions of the intricacies and variations of nest building testify to the artists' absorption in their subjects, while Howard Jones explains that eggs were the centers of birds' existence, "guarded with their lives."[11]

The Jones family made its own publishing opportunity by issuing plates in parts in the older subscription method. The warm praise from the leading ornithologist and bibliographer Elliott Coues, who likened the work to *Birds of America*, implied many in the ornithological community also admired it, despite its modest total of forty-four subscribers. Morse's publication similarly impressed the small community of malacologists but, alas, probably sold fewer than forty copies in the first two years. Fortunately, local catalogs and their creators had other outlets for publication; by the 1870s and 1880s new periodicals for nonprofessional naturalists provided some of these outlets. Atkins and Kumlien published their observations, for example, in local newspapers and in more widely distributed magazines like *The American Naturalist*, *Forest and Stream*, and the *Oologist*. Magazines such as these and the *Nautilus* for malacologists kept the amateur contribution of observation and species enumeration alive into the twentieth century, often being referred to in other scientific treatises.[12]

The dedication and passion of the Jones family and others stemmed from the uniquely "focused" view of the naturalist. The concentration on one division of animate creation, whether invertebrates, birds, or plants, in a particular place yielded endless wonders to those who observed over the years. Each region, when studied so intently, could reveal new species (the Baird brothers, for instance, discovered two new flycatchers in their tiny area) and fascinating habits such as the atypical sexual relations of the phalaropes as noted by Kumlien. Writers claimed that the part of the world under their eye was blessed, as Curtis had boasted of Wilmington's plant bounty. Curtis, on first viewing Wilmington, had thought it a sandy and barren place; his observation rendered it plenteous.

The State Works

The centuries-old linkage of a geopolitical entity to its flora and fauna continued in private and public publications. When state governments commissioned surveys to explore the potential wealth of their natural resources, they often included funds for natural histories of birds, mammals, invertebrates, and plants. The assumption that all (or as many as possible) state plants and animals should be cataloged thus ensured more natural history books. New York's publication was so lavish and extensive as to rival the illustrated monographs.

97

Nature
Writ Large
and Small:
Local, State,
and National
Natural
Histories,
1850–1875

The Private Works

States remained closely identified with their flora and fauna. Zadock Thompson, for example, placed the "natural history" portion of his *History of Vermont* (1842), a good one-third of the work, in front of his "civil" history and gazetteer. Thompson claimed to supersede the catalog in Samuel Williams's 1809 *History* thanks to his adaptation of Nuttall's *Manual of Ornithology* and the Massachusetts natural history survey reports begun in the 1830s (botanist William Oakes contributed the list of Vermont plants) and the addition of his copious "original observations." Indeed, Thompson charts his records of the first day the robin and song sparrow were seen and when the red plum blossomed for the years 1828 to 1841.[13]

As in the era of the colonial explorers, natural history could promote a new area's attractiveness. Increase Lapham, who became one of Wisconsin's early civic and business leaders as well as a respected ethnologist, included flora and fauna in his 1846 book lauding Wisconsin when it was still a territory. He implies that its ample natural resources complement its burgeoning population and stable civic structures and would only support its claims to statehood. His recording of the large game animals and fish assures "the scientific naturalist, the sportsman, and the angler" that "Wisconsin affords a very interesting and highly attractive field." A versatile naturalist, Lapham exerted himself to enumerate over ninety plants in both Latinate and common names and their specific locales in two double-column pages. This list, probably not read item by item by most readers but understood as impartial scientific evidence, adds credence to Lapham's claim that "Wisconsin is as rich in plants as other States in the same latitude."[14]

State Publications

Natural histories "of place" applied to various regions multiplied, thanks to the establishment of geological surveys initiated by most states between the 1830s and 1880s. Many such surveys, in addition to the mapping of geological structures, gave ambitious instructions to the participants to collect specimens and write catalog descriptions. For example, Massachusetts and Ohio commissioned naturalists to compose accounts of mammals, birds, fish, plants, and mollusks; these were published in the 1830s and 1840s. Many of the subjects effected positive and negative economic impact—the trees in George B. Emerson's Massachusetts report, the edible mollusks in Augustus A. Gould's invertebrate study, and insects injurious to crops by Thaddeus W. Harris, for example. Yet these book-length works included species of no apparent economic value, owing to the charge of total inclusion within a geopolitical region.[15]

Because distribution of the hundreds of copies involved out-of-state libraries, schools, and booksellers, these works became important vehicles for natural history outside the states' borders. Contemporary publications such as those of Audubon and Bachman cited Jared Kirtland's works on mammals and birds of Ohio, and Augustus A. Gould aimed for his work on Massachusetts species to become an original contribution to malacology; as he told his "brother conchologist" S. S. Haldeman after three years of effort: "I have taken great pains to compare our marine shells with analogous ones from the other side of the Atlantic—have figured them all—and redescribed them all. I number now about 250 species, nearly one half of which neither Mr. Say or Mr. [Thomas] Conrad [another conchologist] have probably never seen."[16]

The New York Experience

The most ambitious state-mandated natural history work was that of New York in the 1840s. John Torrey listed briefly the physical characteristics, synonymy, and favorite habitats of over fifteen hundred species adapted from his monumental *Flora of North America* in the two-volume *Flora of the State of New York*, with 161 plates. James DeKay covered over eleven hundred species of birds, mammals, fish, and mollusks in his six-part, five-volume *Zoology of New York*, with over 350 plates. As DeKay had little time to observe in the field, he relied on specimens in the New York Lyceum and other authors' work, such as Audubon's *Birds* and Achilles Valenciennes's French publication on fishes, *Histoires des Poissons*. In his ambitious quest to cover complete divisions of the animal kingdom, he included "extra-limital" species outside New York state, such as the Florida manatee. Unlike most of the other less well funded state surveys, the volumes contained ample illustrations originally drawn by J. W. Hill, Agnes Mitchell, Elizabeth Pooley, and William Swinton, in mostly lithographed form. After legislative battles over who would receive the reports, which cost tens of thousands of dollars to issue (lawmakers did not want the expensive complimentary copies sitting unused on legislators' shelves as other reports had), schools, colleges, and private individuals obtained copies throughout the state, and private publishers arranged to sell the natural history section (including the extensive paleontology volumes) separately nationwide.[17]

99

Nature
Writ Large
and Small:
Local, State,
and National
Natural
Histories,
1850–1875

The coverage of so many species, most not limited to New York, made the zoological portion a de facto national monograph in lieu of any other comprehensive catalog before the 1850s. Susan Fenimore Cooper's *Rural Hours* (1851) frequently refers to DeKay's mammal descriptions in her wildlife studies; in a more critical tone, South Carolinian William Elliott corrected his entry on "the devil fish." Henry William Herbert, who wrote sporting literature under the pseudonym Frank Forester, copied DeKay's lake trout text but deplored "the atrociously executed" plates "that for matters of scientific examination . . . are all but useless." The derivative nature of DeKay's text and the generous funding he received from the legislature deeply stirred such fellow naturalists as John Bachman, who with Audubon was searching for subscriptions to the *Quadrupeds,* still in its formative stages: "About DeKay & his book that cost $130,000 . . . By George it is a windy book: the Zoology of my dear native state—methinks the very quadrupeds will cry out murder . . . the Manitoe [*sic*] will rear its head in the rivers of Florida & implore in the name of extra-limital to be let alone."[18]

Other states after the Civil War combined geological surveys with natural history efforts (the huge plant list of Moses Ashley Curtis was made for one such North Carolina survey in 1867), but because of discipline specialization and the successes of the past catalogs, the dual practice died out. Massachusetts revived the Harris and Gould reports by commissioning the entomologist A. S. Packard and the conchologist William Green Binney (son of Amos Binney), respectively, to annotate the texts and by hiring Henry Marsh to execute exquisitely shimmering wood engravings. However, concerns about injurious insects and the economic value of birds did not fade from the state publications, but rather intensified in the annual reports issued by entomological bureaus and agricultural departments, which formed from the 1870s onward.[19]

The Federal Works

Natural history activities had been associated with federal explorations of new territory since the Lewis and Clark expedition collected mammal and bird specimens in 1804 to 1806. Thomas Say and Titian Peale had acted as zoologist and natural history illustrator, respec-

100

THE BOOK
OF NATURE
Natural History
in the
United States

tively, on the 1819 Long Expedition to the Rocky Mountains. Under the supervision of the United States Corps of Topographical Engineers, surveyors continued to travel, map, and record their observations of the rapid expansion of lands ceded to the United States in the Pacific Northwest and the Southwest until 1863, when the Corps dissolved to be reformed as the Army Corps of Engineers.

Publication of the expedition reports meant a boon for natural history. Government funds underwrote hefty reports complete with illustrations. The nation's top naturalists and artists collaborated on the projects under the supervision of Spencer Fullerton Baird, the former Cumberland County, Pennsylvania, naturalist, now assistant secretary of the Smithsonian. Under Baird's guidance, the reports resembled previous national monographs in their scope, thoroughness, and iconography. In a break from their former aversion to theorizing, naturalists who experienced collecting and describing specimens across an unprecedented range developed new ideas relating to the origin of species.

The Process

John Frémont, William Emory, John William Abert, Lorenzo Sitgreaves, Howard Stansbury, and no fewer than six different teams on the Pacific Railroad expeditions (1851–1854) led explorations that included intensive natural history collecting, often under treacherous circumstances. The young naturalist Dr. S. W. Woodhouse, on the 1851 Sitgreaves trip to the Zuni and Colorado Rivers, "received an arrow through the leg, fortunately without doing him much injury" (as he indeed contributed over thirty new botanical specimens and wrote many zoological descriptions for the final report). J. W. Abert persevered in collecting specimens on his sickbed during an expedition to New Mexico, because "gentlemen of the fort would daily visit my room, bringing rare plants and minerals."[20]

In a well-prepared expedition, participants were expected to collect specimens. Plants were placed between papers in a press for drying, then stored in a portfolio (in a pinch, newspapers proving adequate and available); one set of instructions requested, "if the stem is too long, double it or cut it into lengths . . . in the small specimens, collect the entire plant so as to show the root." Smaller mammals and invertebrates were placed in bottles or metal containers filled with alcohol "supplied with tartar emetic [that] will remove any temptation to drinking it on the part of unscrupulous persons"; birds and larger mammals were skinned. Documentation of location and habits was highly prized, as the plants were to be numbered and

each listed with its "day of the month, locality, size, and character of
the plant, color of flower, fruit, etc.," and the eggs of birds were to
be complemented ideally with a skin of the parent bird. As Abert's
bedside experience indicates, all members of an expedition were on
the lookout for new plants and animals. Even the leaders, despite
their responsibilities for managing and completing survey measurements, actively collected. John Frémont and William Emory, for example, displayed a knack for discovering cacti and other plants, and
Emory faithfully wrote a catalog of plants collected daily.[21]

101

Nature
Writ Large
and Small:
Local, State,
and National
Natural
Histories,
1850–1875

The Publications

Before 1850, expedition leaders were only sporadically successful
in publishing these natural history results. The early expeditions,
such as the Long Expedition, had to rely on private publications to
disperse new species discoveries, but government officials, attempting
to match the publicly funded and beautifully illustrated publications
of British and French expeditions, began to grant considerable mon-
ies, as they did for the Great Exploring Expedition of 1838. Those
ambitious volumes limped to conclusion fifteen years after the event
because of fighting among the commander, the scientists, and congressional leaders. The arrival of naturalist Spencer Fullerton Baird
to the second-highest position at the newly formed Smithsonian Institution in 1850 was to ensure complete and relatively timely publications of the army expeditions with an undeniable emphasis on
natural history.[22]

Baird not only took control of the collections gathered in the
expeditions (and often suggested medical personnel or other associates with natural history interests to accompany them), but also supervised the writing, printing, and illustration processes from the 1850
Stansbury expedition to after the Civil War, when other expedition
publications followed the format he had set up. Under Baird, the
natural history sections, as separate sections from the maps, topographical descriptions, and commanders' narratives of the journey
(which also mentioned the new plants and animals observed), took
up a substantial portion of the total text and illustrations. They were
composed as lists of specimens collected over the specific routes
(often supplemented by other collectors in the same region). As in
the private monographs, physical descriptions employing specialized
terminology (such as "areolae crowded . . . berry clavate" in one
cactus definition) and extensive synonymy, followed by general comments on location and habits, predominated. In some cases, such as
that of Robert Ridgway (Baird's own protégé, who accompanied the

102

THE BOOK
OF NATURE
Natural History
in the
United States

1867 geological survey of the fortieth parallel under Clarence King), the field participants wrote the published descriptions, but in many cases the commanders or Baird passed the specimens to the top U.S. naturalists in the speciality, such as botanists John Torrey and Asa Gray and invertebrate specialists S. S. Haldeman and A. E. Verrill. In the monumental *Report of the United States and Mexican Boundary Survey* and *Pacific Railroad Reports,* Baird himself with Smithsonian colleague Charles Girard wrote the bird, mammal, reptile, and fish sections, with the innovation of tables displaying the measurements, locale, and collector of each numbered specimen. Formerly, monographs gave only one set of measurements; now interested readers could compare sizes, speculate about the differences in locales, and know the name of the person who plucked or hunted the specimen. Baird obviously prided himself on comparatively speedy release of the publications in contrast to the ill-fated Great Exploring Expedition, for he willingly explained that his haste was responsible for the typographical discrepancies in the Pacific Railroad reports: "The work was necessarily passed through the press with a rapidity probably unexampled in the history of natural history . . . nearly 400 pages having been set up, read, and printed during the first half of July alone."[23]

Later historians have complained that the Bairdian taxonomic descriptions are dull (the writer and ornithologist Robert Henry Welker bemoaned the "general dead-levelness common to government publications"), but those descriptions by naturalists who saw the specimens in the field offer occasional flashes of graphic detail and personal observation. Woodhouse described the "truly elegant" scaly partridge, attributing to it the characteristic of familiarity with strangers, based on his experience: "It is by no means a shy bird, frequently coming about the houses. I have often observed the male birds perched on the top of a high bush, uttering their peculiar, and, I might say, mournful call." The peccary's flesh was edible, "but it is necessary, immediately upon its being killed, to remove the gland from off the back, which emits a disagreeable odor." Robert Ridgway grew almost lyrical in his praise of the white-bellied swallow's graceful motion, "while each movement caused the sunlight to glance from their burnished backs of lustrous steel-blue, with which the snowy white of the their breasts contrasted so strikingly." The bird passages in J. W. Abert's reports, much in the rhetorical style of Audubon, waxed so poetical that Richard H. Kern, himself an expeditionary artist, confided to the leading American ethnologist, Samuel G. Morton: "One must come to the conclusion that his mind never grasped any idea beyond that connected with 'the sweet warbler of the woods.'

The insipidity and trashiness in his report would make him an excellent contractor for the furnishing of milk man's milk." Budding ornithologist Henry Wetherbee Henshaw lamented that his descriptions in the Wheeler report of the regions west of the one-hundredth meridian fell well short of "complete biographies" but explained that "the character of a topographical survey necessitates frequent changes" and loss of study time devoted to individual species.[24]

103

Nature
Writ Large
and Small:
Local, State,
and National
Natural
Histories,
1850–1875

The plates greatly resemble those produced for the private monographs. Their iconography matches that developed for the disciplines in the last "generation": the birds and mammals are posed as living creatures in habitats suggested by ground and trees, as in the work of Audubon; the reptiles and fish are isolated from the environment with more floating details, as in Holbrook's *Herpetology*; and the plants are shown with ample dissected details, as in the works of Gray. Investigation into the making of the plates indicates that the very same naturalists and artists involved with national monographs were creating the government products. The lithography firms of Bowen & Co., Duval & Co., and Thomas Sinclair, all of Philadelphia, and the Washington firm of William Dougal manufactured the plates for most of the expedition reports in the same style as they were producing private scientific works.

For example, botanist John Torrey guided his protégé Arthur Schott, a young topographical engineer who served on several government expeditions as a surveyor, in Schott's drawing of plant specimens from Stansbury's expedition to the Great Salt Lake (Illustration 3.2). Schott's rendition of a flowering plant he collected gives the impression of a freshly plucked specimen bent to fit onto the drying paper: the leaves still weave about the stem. As Torrey had suggested to Stansbury, the drawings were sent to Joseph Prestele in Ebenezer, New York, to be lithographed on stone and then were shipped to Ackermann & Co. of New York City for printing. The complicated transportation arrangement is reminiscent of the making of Torrey's and Gray's privately published botany illustrations. Gray also probably used his old draughtsman Isaac Sprague and engraver Prestele for his Pacific Railroad contributions.[25]

The botanist George Engelmann, who wrote the cacti monographs for six reports, instructed Paulus Roetter of his town of St. Louis to draw the specimens and then used a variety of engravers, including William Dougal of Washington and five Parisians whom he contacted himself during his travels in Europe. (No fewer than seventy-six plates in the *Report of the United States and Mexican Boundary Survey* are devoted to Engelmann's specimens.) Engelmann favored

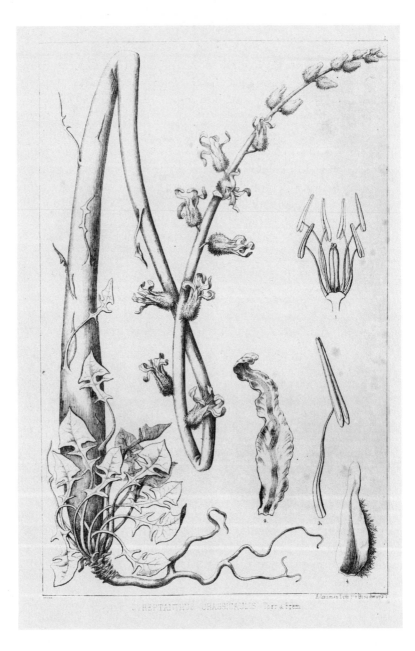

STREPTANTHUS CRASSICAULIS Torr & Frem

Illus. 3.2. *"Streptanthus crassicaulis."* Lithograph by James Ackermann after Arthur Schott, for Howard Stansbury, *An Expedition to the Valley of the Great Salt Lake of Utah* (Philadelphia: Grambo & Co., 1852), opp. p. 384. (Courtesy of the Winterthur Library, Printed Book and Periodical Collection.)

the Europeans' finesse, but the plates engraved by Dougal also convey
widely differing textures from the soft aureole to spiky spines (Illustration 3.3).

Dougal as well translates the seemingly infinite number of smooth scales and raised bumps on the beaded lizard's skin from the Mexican boundary report (Illustration 3.4). The draftsman John H. Richard arranges the creature to display the entire tail and four claws while creating the illusion of the live reptile by skillful placement of the eye highlight and energetically curving profile. (Schott, who had his eye out for all natural phenomena, obtained the specimen.) Such a lifelike impression had become a convention, as in this particular case Richard could work only from a dead example, not the live lizard moving in the southwestern desert. Baird relied heavily on Richard, the snake and frog master of Holbrook's *Herpetology*, but at a distance since Richard lived in Philadelphia and Cambridge, Massachusetts.[26]

105

Nature
Writ Large
and Small:
Local, State,
and National
Natural
Histories,
1850–1875

Illus. 3.3. *"Cereus dasyacanthus."* Engraving on steel by William Dougal after Paulus Roetter, for Emory, *Mexican Boundary Survey*, vol. 2, Plate 39, *Cactaceae* section. (Courtesy of the Winterthur Library, Printed Book and Periodical Collection.)

The vast numbers of plates and unprecedented financial support separates these endeavors from those of private publications. Most government reports were issued in the thousands rather than hundreds, with the Pacific Railroad reports editions and reprints probably adding up to over fifty-three thousand copies of the final volume. As in the case of the New York state survey, economic factors did not seriously compromise quality consideration; Baird often chose more expensive copperplate over lithography technique in certain instances, as in the display of reptile details. Because there was no economy of scale in hand-coloring (for example, Richard consistently charged five cents), not all copies of plates were colored, although the birds and mammals were more likely to be colored than the invertebrates and plants. As Ann Shelby Blum has commented, Baird conservatively favored the already formed methods of representation and production as the best available and most expedient, and so he consistently geared the contracts in that direction.[27]

The huge allotments of money for the total publications (ten thousand dollars for smaller publications such as J. W. Abert's *Reports*,

Illus. 3.4. *"Heloderma horridum"* [Beaded Lizard]. Engraving on steel by William Dougal after John H. Richard, for Emory, *Mexican Boundary Survey*, vol. 2, Plate 26, Reptile section. (Courtesy of the Winterthur Library, Printed Book and Periodical Collection.)

over a million dollars for the Pacific Railroad reports) could relatively easily absorb payments such as the ten dollars for individual drawings, compared to the limited capital of private publishers. Little wonder that over two hundred plates featuring different botanical and zoological species could be funded in both the Pacific Railroad and Mexican boundary expedition volumes—in fact, many more than the landscape scenes that have received the most recent critical attention. Interestingly, reviewers and congressional critics of lavish spending did not specifically criticize the number of plates and amount of text on natural history or favor the potentially more popular landscapes, but all involved implicitly agreed that defining natural productions should remain a major part of the surveys' practice and productions. In a few cases, such as the lake trout in the Salt Lake and the destructive locusts in the Rocky Mountains featured in the Wheeler reports, texts and multiple plates accentuated species of economic importance, but for the most part, any new species received attention. The naturalist's scope of total knowledge superseded, or at least blended with, utilitarian concerns.[28]

The response to the government reports may best be described as mixed. The bulky reports were not designed for easy reading or reference. For example, the zoological and botanical sections of the Pacific Railroad reports are scattered among four volumes and lack consistent indexing and pagination. This inaccessibility provoked a leading English botanist, William Hooker, to complain to his colleague Asa Gray in the letter quoted at this chapter's beginning and caused the western historian William H. Goetzmann to characterize the Pacific Railroad reports as "like the country they were intended to describe: trackless, forbidding, and often nearly incomprehensible." Nonetheless, because of the sheer numbers produced and the policy to provide free copies to legislators, public institutions, and university libraries, the works were disseminated. Some private individuals interested in ornithology could buy the reports from booksellers as late as 1869, when the nascent ornithologist William Brewster purchased a copy of volume 8 of the Pacific Railroad reports in Cambridge.[29]

Most natural history reports, in fact, were issued later under private publishers, indicating the perceived interest in the topics and the inadequate marketing of the originals. Emory's, Abert's, and Stansbury's reports were sold soon after the government issue in new editions from Philadelphia publishers (the interest in the controversial Mormon way of life fueled the Stansbury publication, so Torrey's drier botanical descriptions were dropped). Baird combined his

108

THE BOOK
OF NATURE
Natural History
in the
United States

mammalogy and ornithology sections in the Mexican boundary and Pacific Railroad survey reports into privately published *Birds* and *Mammals of America*, which contain the same lithographed and engraved plates. Two surgeon-naturalists of the Pacific Railroad expedition, George Suckley and James Graham Cooper, consolidated their texts into the influential 1859 *A Natural History of the Washington Territory*.[30]

Private and Public: The Smithsonian Contribution

Spencer Fullerton Baird further contributed to natural history publications by promoting individual monographs in the new Smithsonian publications programs. Baird's love of natural history, manifested by the list of birds that he and his brother had published in their youth, ensured that the Smithsonian Institution, incorporated in 1846, would promote zoology and botany in its larger charge of "increase and diffusion of knowledge." The Smithsonian, in the person of Baird, encouraged private individuals to donate specimens and observations; he and other naturalists coordinated and published the contributions thanks to government funding.

The Publications

After taking up duties as the assistant secretary in 1850, Baird personally supervised the series of illustrated monographs by private scientists, *Smithsonian Contributions to Knowledge*, that would not have been published otherwise. Natural history, as in the case of John Torrey's description of the plants gathered by John Frémont, were thus represented amidst the ethnological and physics essays. More important, Baird, supported by his superior and the Smithsonian director, Joseph Henry, wished the institution to enable the completion of nationwide cataloging of species, not only in the well-established birds and mammals but especially in the less-studied invertebrates. The series of smaller-scale, less expensively illustrated works, the Smithsonian *Miscellaneous Collections*, held these voluminous catalogs.

Perceiving the need for a new catalog of freshwater and terrestrial mollusks in the early 1850s, for example, Baird and Henry accordingly asked their network of correspondents, both private and those attached to military expeditions, to send specimens to the Smithsonian. Having access to the burgeoning number of specimens sent to Baird, the leading American conchologists, William Green Binney (son of Amos), George W. Tryon Jr., Temple Prime, and Thomas Bland, contributed to five installments on the varied classes, entitled

The Land and Fresh Water Shells of the United States, the descriptions derived from their collections and the Smithsonian contributions from across the nation. Continuing the private-public partnership, Binney wrote for the United States National Museum *Bulletin* series (again sponsored by the Smithsonian) *A Manual of American Land Shells*, in 1885, which was based in large part on specimens gathered earlier under Smithsonian auspices. He carefully tabulated the specimens' locations and collectors, as had Baird and the other ornithologists, so the reader could know the geography, number of specimens, and even individual collectors.[31]

Baird and Charles Girard, his Smithsonian colleague, took charge of herpetology and ichthyology. They classified and listed the specimens sent to them in their catalogs of fishes and reptiles published in *Miscellaneous Collections* and in separate pamphlets. The Smithsonian had advertised for North American fish specimens so that Louis Agassiz, the celebrated transplanted European naturalist, could write a more comprehensive monograph. However, Agassiz, a formidable fundraiser, decided to publish his works without Smithsonian aid.[32]

Even the Smithsonian's resources could not fund every national monograph. For example, Thomas Brewer's mammoth work on every egg in the United States, with its beautiful lithographs, was curtailed owing to lack of funds, as was another ambitious venture—the *Trees of North America*, written by Gray and illustrated by Sprague and Prestele. Baird further influenced many contemporary private monographs in his lending Smithsonian specimens and freely giving advice on publication and zoological nomenclature. It was he who advised William Henry Edwards on the title for his butterfly monograph, *Butterflies of North America*, discussed later in this chapter, and who sent Daniel Giraud Elliot many birds featured in the latter's beautifully lithographed grouse book. Baird's prompt replies to Elliot's extensive requests earned him (quite justly) the dedication in the work.[33]

Collectors across the Nation

The Smithsonian from the early 1850s had spread the request for nationwide specimens through circulars (one specifically addressed to Hudson's Bay Company employees, from whom Baird eventually obtained valuable northern items) and by word of mouth. Some Americans not especially interested in natural history, in a spirit of helpfulness to a national organization, sent in interesting specimens if they had heard about the Smithsonian's request, as in the case of the many fishermen who excitedly contributed specimens

110

THE BOOK
OF NATURE
Natural History
in the
United States

to the Agassiz project. Other contributors had natural history expertise or developed, through the experience of contributing to the
Smithsonian, interest and skill in the discipline. The botanist Moses
Ashley Curtis and his son sent mouse skeletons through the mail; they
packed them so that the stamp at the top corner would not damage
the specimen. Benjamin Wailes, the Mississippi College president
who wrote that state's natural history survey in 1854, sent in reptiles
and shells. Doctor P. R. Hoy of Racine, Wisconsin, similarly wrote
articles on his state's flora and fauna and built an extensive collection
of local birds and shells; one of his Smithsonian contributions was a
life history of the *Amblystoma luridum,* an orange and bluish salamander "found about my cellar after a wet night." Trapping the individuals evidently demonstrated to him that "they resent any insult offered
to their mouth or eyes by quick and repeated strokes with their ample
tails."[34]

Among the most active contributors of reptiles in the early 1850s
were Charlotte Paine and Mary Daniels of western North Carolina,
who had fourteen of their specimens described in the Smithsonian
reptile catalog. Daniels so retained her interest in collecting species
that after she moved to a new location she offered Baird to "build a
collection" for the Smithsonian from local contacts she was forming,
because "snakes, lizards, and frogs were pouring in from around."
The Reverend Charles Mann of Gloucester County, Virginia, sent
Baird the salamanders and eggs his sons had found (and also requested in return some specialty seeds from the Patent Office). In
order to encourage collectors to persevere and become more informed, Baird sent them the Smithsonian catalogs and instructions
for preserving specimens, and, in the example of Valerie Blaney,
named a snake after her. G. W. Binney particularly acknowledged in
his works valuable helpers such as Mrs. George Andrews of Mitchell
County, North Carolina, "to whom we are indebted for our knowledge of the richness in molluscan life of this and other mountains of
the region," and Miss Annie Law, for her years of searching for land
snails wherever she lived, whether in Tennessee, Georgia, or California: "Once she took a journey of several weeks, by wagon, over mountainous roads, to the locale where *Vitrinizonites* was originally found,
in search of the living animal, which she kindly sent to me, and thus
fixed the generic character of the species." Law had served as a
schoolteacher of botany and music and maintained correspondence
with another conchologist, James Lewis, who with Binney named after
her new species she had sent to them.[35]

The number of these dedicated collectors is minuscule (twenty-

six private individuals sent specimens described in the 1853 catalog
of reptiles, and Binney refers to a handful of correspondents) compared to the general population, yet their influence was inordinate.
The ability to obtain and compare individuals from one species over
a geographical area helped to distinguish its range and, as the next
section discusses, led to greater precision in ascertaining varieties and
subspecies. Collectors in California, Texas, and the North, however
few, gave enough material for new species to be discovered. Even
though he admitted his collections were imperfect, Binney thought
his knowledge of the Pacific, Central, and Atlantic regions adequate
to discuss the general distribution of genera throughout the North
American continent, thanks to far-flung correspondents. Tangible rewards such as published acknowledgments and gifts of books awaited
the faithful correspondents, yet the hours spent in often unpleasant
conditions or in delicate and time-consuming tasks such as blowing
out eggs, skinning birds, or pinning insects suggest additional motivations—those of intellectual satisfaction and pride in the achievement
of discerning and sometimes discovering species.[36]

Nature
Writ Large
and Small:
Local, State,
and National
Natural
Histories,
1850–1875

The Private National Monographs

Private individuals still pursued the grand ambition of publishing
all U.S. species of one branch of natural history. They, like the previous generation, wished to equal the Europeans in substance and style.
Their illustrations continued to rely on fine engraving and handcoloring, increasingly rarefied processes compared to the rapidly developing form of steam-driven chromolithography. Personal achievements over a lifetime plus contributions from correspondents across
the United States made up the original material of the works. These
personal efforts constitute a last hurrah of the all-encompassing national goal. The increase in known species made completion of such
a project unlikely, and changes within the disciplines turned scientists
from fieldwork and cataloging to other priorities, such as physiology
and laboratory studies.[37]

These authors expanded on the "inhabitant" innovations. The
life history format proved a satisfactory vehicle for new discoveries
still occurring in the West, corrections of previous accounts (now
some American), and personal observations. The iconography of the
species in characteristic attitude inspired ornithological artists, and
entomology reinvigorated its conventions in the unique monograph,
William Henry Edwards's *Butterflies of North America.*

Sullivant's Mosses

Ohioan William Starling Sullivant and his second wife, Eliza, had taken up botany in their adults lives, focusing on the unusual speciality of nonflowering plants, particularly mosses. Both became adept at dissecting and drawing the minute distinctive parts under the microscope, and he corresponded and exchanged specimens with the top American and European botanists. By the 1850s he had published the extraordinary *Musci Alleghanienses*, in which actual dried specimens of the mosses acted as illustrations, and contributed descriptions to Asa Gray's botany manuals.[38]

Sullivant meanwhile was dreaming of an even more "formidable work"—in his words, "a Bryologia Americana . . . with full descriptions & figures of every species I can lay hands on." After Eliza died, he trained "a poor Prussian" artist, August Schrader, to perform dissections and allowed him to live on his estate. Sullivant published his *Icones Muscorum*, an imperial octavo work "of those Mosses peculiar to Eastern North America which have not been heretofore figured" with 129 copperplates, in 1864. William Dougal, veteran printmaker of many government surveys, engraved Schrader's drawings, as Sullivant eschewed lithography for the clarity and fineness of line some naturalists thought only copperplate could provide. Gray in his obituary remarked on Sullivant's devotion to his project, having invested over six thousand dollars of his wealth in support of Schrader and production of the work. Sullivant's lifelong "ideal," according to Gray, "was no other than to perpetuate all the mosses of the United States in clearly drawn illustrations."[39]

Morgan's Beaver

Lewis Henry Morgan's *The American Beaver: Its Life and Works* (1869) may be the most uniquely focused monograph in that it is devoted to a single species. Yet Morgan (1818–1881)—who as a young man so treasured his copy of the New York state survey report that he bequeathed it specifically to his future wife in a memorandum—is now best known for his ethnological writings inspired by the Iroquois and his influence on anthropology, not for his natural history interests. As one of the directors of a railroad system designed to carry mineral ore from the lower slopes of Lake Michigan to New York, he journeyed every summer through "the most remarkable beaver district" near the southwest shore of Lake Superior and, fascinated by beaver dams, lodges, and meadows (flats near rivers in which the flooding created by the dams drowned trees), forsook his trout-fishing for "the beaver." He took the opportunities of Rocky Moun-

tain travel to compare beaver architecture, and in the early 1860s he
and loyal friends—who luckily were amateur photographers—
recorded several lodges and canals after first cutting down trees to
clear the view. Friends sent many skulls (one joked that the trappers'
habits of hanging up a carcass after the pelt had been taken made
obtaining sixty some skulls remarkably easy), and a doctor friend performed a dissection to ascertain differences between European and
American species. (Morgan believed they represented separate species but presented measurements so readers could make their own
decisions.) Morgan compiled so much material "year after year" that
he decided to publish it, because "my restrained curiosity has cost
me a good deal of time and labor."[40]

The fruit of his absorption joins his original empirical observations with generalizations on anatomy and habits derived mostly from
Cuvier and other European authors, "although the author desires to
avoid compilation" of quotations from previously published works.
He devotes chapters to anatomy but gives most attention to describing the huge dams, lodges, and canals he had watched, measured,
and photographed. He "regretted" breaking through a lodge to photograph it but "was surprised and gratified to find that the lodge had
been completely restored by the beavers" by the following year, and
he also lay in wait at night to see a wary group of beavers rebuild after
railroad tracks cut through their homes.[41]

The variety of structures and their adaptations to changing environs led Morgan to comment on the familiar topic of animal intelligence in terms often reminiscent of earlier musings on the great
similarities of human intellect and animal instinct. "In choosing the
sites of their lodges, so as to be assured of water in their entrances
and at their places of exit, too deep to be frozen to the bottom; in
the adjustment of the floors of their chambers to the level of the
ponds; and in their appreciation of the causes of a change of level in
these ponds, as well as of the remedy, decisive evidence seems to be
furnished of their possession of a *free intelligence*, as well as of construc-
tive skill." Canals offered more evidence favoring intelligence over
instinct: "In the excavation of artificial canals as a means for trans-
porting their wood by water to their lodges, we discover, as it seems
to me, the highest act of intelligence and knowledge performed by
beavers . . . to conceive and execute such a design presupposes a
more complicated and extended process of reasoning than that required for the construction of a dam." His concluding chapter, "Animal Psychology," generalizes about humankind's relationship,
particularly its responsibility, to animals in arguments influenced by

113

Nature
Writ Large
and Small:
Local, State,
and National
Natural
Histories,
1850–1875

114

THE BOOK
OF NATURE
Natural History
in the
United States

the animal humanitarian thinkers of generations past, such as William Bartram. Morgan hints that awareness of "the mute possession of a thinking, and reasoning, and perhaps an immortal principle" in all animals may stem the human excesses of carnage beyond "legitimate wants." Without such self-constraint, as civilization increases, animals seem doomed to "extirpation or domestication."[42]

Continuing the Birds

Ornithological monographs directly followed the path of Audubon, down to the choice of lithographer. John Cassin designed his *Illustrations of the Birds of California, Texas, Oregon, British and Russian America* (1856) in the style of the octavo *Birds of America*, with the same lithography firm of Bowen & Company of Philadelphia translating George G. White's and William Hitchcock's drawings of specimens from western environs not figured in Audubon. Charles Maynard's *Birds of Florida* (1872) similarly offered subscribers parts with octavo hand-colored plates. Daniel Giraud Elliot and other artists in his *New and Unfigured Birds of the United States* (1869) drew birds including falcons and turkeys life-size so that Bowen & Company could lithograph them in a magnificent folio size and thereby "finish" the Audubon double elephant. As in his earlier grouse monograph, Elliot based much of his textual and illustrative descriptions on specimens lent by Baird at the Smithsonian. After finishing the birds in his native lands, Elliot in fact moved to London for ten years and wrote monographs on foreign exotics like the birds of paradise with German-born Joseph Wolf acting as his draftsman.[43] Wolf was regarded by many to be the finest animal artist involved with scientific illustration.

Baird with his young assistant, Robert Ridgway, and older colleague and correspondent of Thure Kumlien, Thomas Brewer, wrote the epic three-volume *A History of North American Birds: Land Birds* (1874), featuring the most extensive life histories of U.S. birds yet published, under the auspices of Little, Brown & Company, the successors to Nuttall's old publishers. Ridgway not only designed the woodcuts but also drew the most important birds, which were lithographed and hand-colored. Lavinia Bowen, who had colored the octavo editions of Audubon's *Birds* and *Quadrupeds,* hand-colored the plates from his models. Ridgway excitedly envisioned the work as the apex of the U.S. tradition: "it will be by far the handsomest and most accurate collections of plates of N. Am. birds ever published." His promises to Baird further indicate that both wished to continue in the visual iconography tradition of the bird in characteristic natural pose against background. Ridgway struggled in one case: "I am still confident of pleasing you with the Robin, for while here I shall make studies of a caged specimen, in its various characteristic attitudes." His intent to "Pay more attention to back-grounds, and study [Joseph] Wolf all the time" demonstrates that he was acutely aware of the new master on the international scene.[44]

115

Nature
Writ Large
and Small:
Local, State,
and National
Natural
Histories,
1850–1875

Surely Baird and Ridgway remembered the precision and magnificence of the Audubonian illustration, particularly as they dealt with the lithography firm employed by the Audubons. They preferred hand-coloring over chromolithography to meet their extremely exacting demands of precise coloration and shading, exhorting their colorist, "I hope these are done with great care." The long-distance correspondence between Baird and Lavinia Bowen, the widow of Audubon's lithographer John T. Bowen, demonstrates that such standards were doomed, perhaps never to be achieved completely (and ironically recalls the correspondence between the Audubons and her husband in which he asserts that he is trying his best). Although generally pleased with the work, he criticized a tone on her catbird and wished for a greenish shade on the red-headed woodpecker's back. Bowen contended that she saw no greenish shade in the woodpecker pattern sent her: "I send you today three more plates of Birds, I have made them as near possible to the Specimens sent by Mr. Ridgway— "The 'Red headed Woodpecker' does not reflect the least tint of green." As she evidently wanted as perfect a work as Baird and Ridgway, she complained that the lithographer had not printed the outlines as clearly as she would have liked. After months of coloring, she decided to raise her prices because of Baird's continual demands, or as she more tactfully expressed, "the style of the works is so very different [from the Audubon octavos] . . . requiring much more care and nicety of coloring in order to give them the finish of the Original Copies . . . it would be impossible for me to do them at the old prices."[45]

Edwards's Butterflies

Entomology, more specifically the study of butterflies, provided the climax of the individual effort at the high-quality, national monograph. William Henry Edwards conceived the idea for his *Butterflies of North America* about 1864 and published its three volumes in 1872, 1884, and 1898. The span of years signifies not only changes in reproductive techniques and publishing firms (the first two volumes featured hand-colored lithographs, the last photolithographed; the first was published by subscription in cooperation with the American

116

THE BOOK
OF NATURE
Natural History
in the
United States

Entomological Society, the others with Houghton Mifflin Company) but also his single-minded determination. Edwards, born and raised in midstate New York, was so knowledgeable in ornithology early in his life that as a Williams College student in 1842 he wrote Audubon to inform him about the chestnut-sided warbler's nest sites, information that Audubon, Nuttall, and Wilson had not known. After becoming one of the first Americans to explore the Amazon River, Edwards shifted into the more prosaic life of New York banker and refocused his natural history interests on butterflies. As director of a coal mining company, he settled in Coalburgh, West Virginia, and was to reconnoiter for lepidoptera in his beloved Kanawha Valley until his death in 1909. When he finished paying for the plates for a monograph on the *Sphingidae* family that stopped production in 1863 before completion, Edwards turned to the larger project of commissioning illustrations for all the many undescribed butterflies of the Rocky Mountains. He wrote to Spencer Fullerton Baird, who introduced him to many collectors, "We have nearly 30 species, many very large sized ones, Californian mostly, and such a work will be handsome as well as useful to entomologists." The scope of the project multiplied in the course of the next four years because Edwards himself "named more than 70 species U.S. & B.A. [British America] butterflies" and noted that he knew over eighty other species, "none of which are anywhere figd." In his 1868 preface to the *Butterflies'* first volume, he admits that the goal of figuring the burgeoning number of species "in such a world as this continent affords" seems impossible yet worth the struggle: "nothing is more perplexing and discouraging to the beginner than dry, unillustrated descriptions."[46]

Like the other illustrated national monographs, or "formidables," to use Sullivant's phrase, Edwards was keenly aware of European standards in theoretical and physical presentation and frankly wished his American work to surpass them. In the letter to Baird about the project's genesis, he mentioned an 1850s English publication, *Hewitson's Exotic Butterflies*, as a model for plate format and constantly refers in his text to the Frenchman Boisduval's hand-colored engraved works on American butterflies, many engraved after drawings by John Abbot and John LeConte. In the tradition of Audubon, he blended knowledge of precedents with his own field research in the standard format of species descriptions juxtaposing taxonomy and life histories, complete with plates, perhaps because Baird had recommended such a text including "localities, habits [to] balance the plates." For example, the uninitiated might not believe invertebrates to possess interesting and characteristic habits, but Edwards's account of his discovery of the female *Diana* bespeaks its special

117

Nature
Writ Large
and Small:
Local, State,
and National
Natural
Histories,
1850–1875

movements, especially as they reveal its relationship to intruders, which he observed in the field: "exceedingly alert and wary species, differing in this from our other *Argynnides*. At the slightest alarm it will fly high into the woods, near which, upon the narrow bottoms or river slopes, it is invariably found."[47]

Edwards also perfected a technique of raising specimens from known species in boxes by enclosing a gravid female on the food plant. Before this development, naturalists like John Abbot usually had little choice but to raise specimens from unknown pupae. With this ability to ascertain the exact parentage of butterflies, he made important discoveries in polymorphic forms (members of the same species that vary in coloring and markings apart from sexual variation). For example, varying forms of one species of the *Ajax* butterfly came from different parentage, often depending on the season; Edwards clarified the taxonomic question of whether these forms are species or varieties.

Edwards's taxonomic discoveries and advances in observing life stages inform the iconography of *Butterflies'* plates. He was fortunate in receiving the longtime services of Lavinia Bowen and her sister, Mrs. E. Leslie, as colorists and of Mary Peart, who drew the original sketches and then transferred them onto lithographic stones. John Cassin, then head of Bowen Company, had introduced Peart, who was to draw some images for the second edition of Gould's *Massachusetts Invertebrates*, to Edwards, who was looking for an artist. As Edwards remembered, she "at the beginning knew nothing of the structure of butterflies. I called her attention to the peculiarities of the small organs, the legs, the antennae, and palpi, and gave her a net with which to take live butterflies in order to make sketches of these organs." The minute distinctions in organs, wing markings, and colors between varieties made Peart's and the Bowen firm's tasks especially challenging, yet Edwards claimed complete satisfaction. He once boasted to Academy of Natural Science members that Peart knew more about butterflies than the specialists, and he paid tribute to Mrs. Bowen in his memoirs. In order to render all the stages of the butterfly life cycle—from egg to several larvae stages to pupae and adult, male and female—Edwards and others sent the eggs and larvae to Peart for her to raise and draw. Accordingly, the plates are often crammed with many larvae as well as the egg and chrysalis. Other contemporary U.S. entomologists proudly noted that the Europeans had no such complete knowledge of early stages, and English reviews said "they [the images] will bear comparison with the best that have been given in iconographical works" (Illustration 3.5).[48]

Edwards depended on wily and dedicated collectors throughout

Illus. 3.5. "Interrogationis." Lithograph by Lavinia Bowen after Mary Peart, for Edwards, *Butterflies of North America,* vol. 1. (Courtesy of the Library, Academy of Natural Sciences, Philadelphia.)

the United States and Canada to send eggs and adults. To cultivate Mrs. Christina Ross, the wife of a Hudson's Bay Company employee who became interested in butterflies after another Baird correspondent on an exploration trip gave her a net and pins, he sent her a book on insects at Baird's suggestion. For her contributions, a butterfly was named in her honor. The connection of West Virginia to the nation via the C&O Railroad made new collectors available to Edwards. "In the next year [1871] all places which could be reached by rail, such as Florida, Texas, Arizona, California, and Colorado, as well as much of British America, were in my reach. Correspondents in all these regions were instructed how to obtain eggs of butterflies, and eggs were sent me through the mails." These correspondents included Annie Wittfield of Georgiana, Florida, and W. G. Wright of San Bernardino, California. One correspondent's devotion and persistence exemplifies these collectors. Theodore Luqueer Mead, "young and enthusiastic," asked where to forage for an abundance and variety of butterflies. Edwards told him of the Kanawha Valley's entomological riches, and so began the many yearly forays ("whenever he appeared entomological work went on finely," according to Edwards) during which Mead collected hundreds of adults and eggs. In Edwards's *Parnassus* description, Mead relates how he found one variety on an expedition to Colorado and "captured forty-one specimens," including many females who later laid about one hundred egg sacks on his enclosures. He supplied Edwards with data on characteristic movement and food plants. The helpful stagecoach rider who stopped the stage for an hour so Mead could collect at a certain altitude on the Continental Divide may deserve a modicum of praise. Mead's intensive study led him to note differences in color at different elevations and to discard confidently the notion that the peculiar pouch of the Parnassus butterfly was formed "in coition": "the period of connection between the sexes of these butterflies [would] be very long. In that case I should have found many pairs in coitu, whereas I have not so found a single pair."[49]

The ample text of the life history format gives full play to Edwards's musing on the wonders on his subject. Natural history had not become so specialized nor emphasis on brevity so great that it excluded his excerpt from Spenser's poetry on the fiercely red horn of the *Libythea Bachmanaii*, a touch reminiscent of the poetical interpolations of Alexander Wilson. His thoughts on the communal workings of tent caterpillars, who enacted specialized tasks, harkens to the wonder of God's creation and musings on the similarities between human and animal intelligence. He never tired of watching the chrysalis emerging and shared his amazement with the reader:

119

Nature
Writ Large
and Small:
Local, State,
and National
Natural
Histories,
1850–1875

120

THE BOOK
OF NATURE
Natural History
in the
United States

Suddenly, and to a looker on without any premonitory symp-
tom, a rent takes place in the skin at the back of the head
just wide enough to allow the passages of the chrysalis, the
head of which at once emerges . . . the whole process is most
interesting to witness and excites renewed wonder with every
repetition . . . how to strip the skin and much more the legs
by which the creature is suspended, without losing its hold,
and at the same time to securely fasten the chrysalis, is a
problem that would seem impossible to solve, and yet this
little insect accomplishes, unerringly.[50]

Development of the Disciplines

The combination of private dreams and experiences plus public
funding proved a potent boost to publications, particularly of the
hitherto understudied invertebrates and reptiles. The naturalists fully
recognized the last generation's contributions, but in the aftermath
of Darwin's theory of evolution they adapted taxonomic concerns to
accommodate the origin of species. American naturalists were not
especially original in pure theoretical thought, but in the process of
intently studying species over a vast continent, they provided striking
examples of Darwinism in action and, in the case of ornithology, contributed nomenclature innovations.

Botany's Riches

The government surveys benefited the discipline of botany in the
abundance and diversity of specimens pouring from the new territories and the ability to publish the new species, especially with illustrations. The surveys' illustrations portraying life-size specimens with
ample details were drawn and produced as Torrey and Gray would
have chosen for their private publications if they could have afforded
them. Torrey's botany section of the Mexican boundary survey alone
included 136 lithographs of plants. The bundles of dried specimens
flooding the offices of Torrey and Gray gave them unique opportunities to study individual species from very diverse geographical regions. Gray's intimate knowledge of Japanese species, brought to him
by the Perry expedition members, and U.S. species (both western and
eastern) allowed him to recognize striking and seemingly inexplica-
ble similarities between Japanese and eastern North American flora.
He posited a theory that the one-time linkage between continents
resulted in shared flora around the Bering Straits, and that this flora
subsequently spread southward owing to glacial cooling. Such a the-

ory helped Darwin to discredit the belief in separate origins of closely related species espoused by Gray's Harvard colleague Louis Agassiz.[51]

121

Nature
Writ Large
and Small:
Local, State,
and National
Natural
Histories,
1850–1875

The intense treatment given to one plant family shows how forms representing the unfamiliar "other" could be familiarized and even domesticated. Europeans had associated cacti, so different from temperate flora, with the New World since the sixteenth century and figured them in map margins. The nineteenth-century army travelers linked the southwestern landscape to the peculiarly American plants throughout their journals, for these plants signified the desert constantly. William Emory frequently noticed the dominant cacti when he came upon new scenery ("Cacti in endless variety and gigantic in size" in one memorable case); he relished his first taste of the prickly pear, with its "flavor of a lemon with crushed sugar." The army men had been instructed specifically to describe, collect, and make diagrams of cactus specimens. Emory proved so effective a naturalist that species he harvested were named in his honor. George Engelmann's separate monographs on cacti, in four reports attached to the government expeditions, gave lengthy taxonomic descriptions. Engelmann had also enhanced his knowledge of their life cycle by growing cacti sent by settlers in the new territories (luckily, live cacti traveled extremely well). Unlike European botanists, who knew only greenhouse cacti that did not produce flowers or fruit, he boasted, "I am now able to distinguish all the different genera of cactaceae by their seed, and sometimes even sections of one genus." His artist, Paulus Roetter, and the engravers rendered every tubercle, spine, blossom, and branch of the cacti, so the unknown became public knowledge (see Illustration 3.3). The topographical artists made the cactus the major protagonists in their views. The cactus became a houseplant during the 1840s, with "Christmas" cacti even appearing on album quilts of the era.[52]

The Rise of the Other Classes

The surveys and the Smithsonian's efforts greatly promoted invertebrate and other non-mammal study. The naturalist's mandate to describe and picture every native species in the surveys included the previously ignored or underrepresented reptiles and fishes, so John H. Richard was commissioned to execute exquisitely detailed fish, textured "horned frogs," and glittering-eyed snakes. Baird wished to catalog classes less studied than birds and mammals, so the Smithsonian became active in collecting mollusks, fishes, and reptiles. Its authors, like Baird and the fish specialist Charles Girard, and such private individuals as William Green Binney issued descriptions based

122

THE BOOK
OF NATURE
Natural History
in the
United States

on collections and observations. Malacology, in particular, continued the tradition of the national catalog, thanks to the Smithsonian's call for freshwater and land snails. The top malacologists teamed to present the multivolume definitive catalogs. In the many installments of *Terrestrial Air-Breathing Mollusks of the United States*, sponsored by the Smithsonian and the Harvard Museum of Comparative Zoology, Binney continued his father's work begun in the 1850s. By the 1885 publication of his *Manual of Land Shells* for the Smithsonian, Binney had knowledge of so many American species across the continent that he could hypothesize how genera and families developed in their discrete geographic regions according to the fossil record and his knowledge of the evolutionary principles derived from Henry Wallace Bates's 1876 treatise, "The Geographical Distribution of Animals."[53]

Government agencies and private individuals further explored insect life histories in state bureaus and the standard monographs. Edwards's close attention in the field clarified taxonomic questions. The Illinois state entomologist Benjamin Dana Walsh, in his study of the gall gnats, described fifteen closely related species, each feeding on different willows. In his view, forcefully put forth in defense of Darwin's theory in the early 1860s, such specialization could only mean multiple forms adapting to specific food plants from one original form.[54]

Beyond Audubon: Bird and Mammal Studies

Ornithology and mammalogy obviously grew from both private and public publications in terms of new species described and also the theoretical issues stemming from national collecting. Cassin and Elliot followed in the high-quality text and plate genre supported by subscriptions, while government publications featured similar imagery often by identical artists and firms. The government publications used the base of Audubon for birds and of Audubon and Bachman for mammals, to which they added further variations of size and color among and within the species, using the tables of measurements charting individual specimens.

Ornithologists also confronted theoretical issues stemming from national collecting. The process of comparing and reporting these differences over vastly varying terrains, latitudes, and altitudes induced Baird, the Harvard University Museum of Comparative Zoology's young star, J. A. Allen, and naturalists with field experience in the West such as Ridgway and Coues to emphasize differences in form due to climatic differences (e.g., darker coats and smaller size at higher altitudes). These variations, when the species became geographically separated, eventually led to further speciation, but before this occurred were best considered subspecies. American ornithologists, forced to reckon with variation more often than their European counterparts, suggested trinomial nomenclature, the addition of a third term to the system in use since Linnaeus, to indicate the subspecies. European ornithologists and the Americans themselves, such as Ridgway, Allen, and Coues, recognized the innovation's national origins by calling it a product of the "American school." One British ornithologist, Alfred Newton, sounds envious when he relates the Americans' new nomenclature to their opportunities to collect many specimens of one species over a continent: "[T]he small size of our own quarter of the globe compared with that of North America, and still more the short series of specimens which existed even in the largest of our collections, forbade the generalizations that at once became possible and almost suggested themselves when the vast aggregations obtained by Baird and the elder Agassiz [at the Museum of Comparative Zoology] were studied and compared." Thus, evolutionary principles underlay the development of climatic variation and were so firmly adopted by ornithologists that Coues in 1879 could claim boldly (in his usual outspoken manner) that all classification systems are "evolutionary," for "today we utterly disregard any . . . that does not proceed upon the understanding that all birds are descended from a common ancestor."[55]

Such changes within the discipline, along with a wariness of the larger-than-life reputation of their immediate predecessor, Audubon, explain the ambivalent attitude of the ornithologists of this generation toward him. They built upon his descriptions, yet soon after his death in 1851 criticized some of his findings as skewed because of careless notetaking and condemned several figures for their "dramatic" postures. Thomas Brewer, in correcting nesting errors, noted that Audubon's "enthusiasm" led him to "conclusions . . . now known to be visionary, but which his exuberant imagination, now and then, led him to put into printed words." Elliott Coues in his bibliography claims more boldly that Audubon "liked to exaggerate and 'embroider,' and make his pages glow like a humming-bird's throat, or like one of his own marvellous pictures" and that William MacGillivray "supplied what was necessary to make his work a contribution to science" with his nomenclature and technical descriptions. These ornithologists, who fully realized other older contributions to their discipline, thought the image of the sole intrepid traveler and great American scientist built by the popular press to be overinflated. (In 1841 a disgruntled S. S. Haldeman privately complained to Augustus

124

THE BOOK
OF NATURE
Natural History
in the
United States

A. Gould that the vaunted traveler had made no journey longer than "a trading expedition.") Coues did not discount the personal element or the place of an eloquent and graphic description in ornithology. In fact, he praised Alexander Wilson's "fire divine" and gruffly paid Thomas Nuttall, another master of the stirring life history, the ultimate compliment: "Nuttall, like fine wine, improves with age and needs none of my bush."[56]

However, the life history, with its usual accompaniment of personal interjection and interpretation, was increasingly separated from scientific species descriptions and was reserved for "popular" ornithology, to be enjoyed by amateurs and non-bird lovers. Coues himself weaves some humorous wordplay and rich similes in his *Birds of the Colorado Valley* (for example, the canon wren passage, "we remember the 'rift within the lute'; in the Canon Wren, we have the lute within the rift—a curious little animated music-box, utterly insignificant in size and appearance, yet fit to make the welkin ring with glee"); he also openly cast technical description out of these life history sections "to meet the tastes and wants of the public." In the decades to come, only highly trained ornithologists, learned in languages and the nascent disciplines of physiology and fossil studies, were encouraged and expected to wade through tables, appreciate climatic differences, and decipher Latinate trinomial nomenclature. Yet honor continued to be paid to those masters of the past, as when the ornithologist and bird artist Robert Ridgway named his only child Audubon.[57]

The Continuing Themes

Composing the national monographs stimulated American naturalists to comment on new theoretical aspects such as evolution and previous concerns like species extinction. Coverage of a vast territory stimulated these developments. The monographs again depended on the artwork of many individuals and the contributions of local observers, now spread across a greater area. The local observers, in their absorption with their homes, made possible the nationwide focus.

Monographs and Evolution

The American contribution to an understanding of the origin of species developed from the older concerns of species identification and life history study. Gray's knowledge of a continent's botany supported his subsequent explanation of the similarities between far-eastern and eastern North American plants. William Green Binney was able to explain how snail genera evolved in reference to geographic locations thanks to the specimens collected in the Far West and South. Ornithologists understood variations within species through climatic variation and refined the concept of subspecies in different locales. The geographical span of the national surveys and their widespread collecting encouraged this investigation of species development.

125

Nature
Writ Large
and Small:
Local, State,
and National
Natural
Histories,
1850–1875

William Edwards's exploration of the varieties of butterflies epitomized the American exposition of Darwinism. A German scientist, August Weismann, who was working out the implications of evolution in his work on polymorphs, hypothesized that the winter or autumnal forms were the "original" species and the warmer weather forms evolved later. Edwards, who corresponded with Weismann, placed boxes of the *Ajax* butterfly forms on ice and reported that the warmer weather forms did revert consistently to cold. Englishman Henry Wallace Bates, an ardent exponent of Darwinism, posited polymorphs that inhabited a wide region suggested evolution in that the varieties physically separated grew apart from one another and the "middle" or transitional forms disappeared. Edwards, using his vast collection of one dimorphic species, *Grapta*, collected across Canada and down to Texas, found such a case. Wallace, Weismann, other American entomologists, and even Darwin praised his efforts as thoughtful supports of the latest theories based on sound fieldwork.[58]

The many terrains and habitats fostered naturalists' insights and made their achievements uniquely American. The continent's physical presence had long held the opportunity of special contributions to European culture; for example, important astronomical observations of the 1769 transit of Venus were possible from various points in the colonies. Now, as Alexander Wilson and other naturalists had predicted, Americans were publishing original contributions based on their experiences in the United States. The Europeans appreciated the interpretations and theories inspired by the physical resources "that almost suggested" (to quote Alfred Newton once more) the American study on climatic variations. The wilderness had promised exceptionalism since the European age of discovery. Now, finally, through the findings of natural historians in a variety of disciplines ironically inspired by an Englishman's theory, American authors met these expectations.

Implications of Expansion

Although the nation's political boundaries failed to adhere consistently to natural or climatic regions, the geopolitical contours en-

126

THE BOOK
OF NATURE
Natural History
in the
United States

couraged naturalists to compare and extend biotas. Surveys of New York and other states welcomed the changes in mountains, plains, and rainfall as indicating the wealth and diversity of their regions. The intense observation of inhabitant naturalists such as Harmon Atkins and Thure Kumlien in the limited areas of Locke County, Michigan, and Lake Koshkonong, Wisconsin, allowed larger comparisons in state and national works. The temptation to include Canada and Mexico proved great, as evinced in Edwards's *Butterflies of North America*, which overran the borders, as had Audubon and Bachman's *Quadrupeds of North America*. Edwards, for example, "needed" many specimens of one genus from the north to demonstrate the formation of species over time and area. The presence of the same species in both nations supported inclusion, yet one suspects some comparatively mild form of imperialism at play. Beautiful southern species such as the iridescent Mexican turkey and brilliant subtropical butterflies tempted Daniel Giraud Elliot and Edwards to portray them. Baird casually advised to Edwards in 1868 that he should include "North America" in the *Butterflies* title, because "One day it will be all United States, and United States will be North America."[59]

The territorial expansion provoked steady commentary over the disappearance of wildlife and the transformation of the wilderness. Lewis Henry Morgan saw the railroad as a personal blessing that introduced the beaver's architecture to him, yet he wished to photograph and record the grand structures before civilization, literally brought by the same railroad, destroyed the builders. The government naturalist on the exploration of the one-hundredth parallel, H. C. Yarrow, evinced deep concern for the economically important lake trout, which was easy to catch: "How long, however, this condition of affairs will last it is impossible to say, the supply having greatly diminished during the past few years, owing to the reckless methods of fishing and increase in the number of fishermen; moreover, a larger demand is now made for this fish, owing to increase in the number of settlers. The decrease in the yield may be roughly estimated at about one-third." In his thorough description, Yarrow bluntly said that the existing laws to prevent overfishing were useless because "no attention is paid to them by some greedy individuals, who think only of filling their own pockets at the expense of future generations." Peter Madsen, "an intelligent Danish fisherman of Utah Lake . . . placed at our disposal data obtained during many years' experience acquired in this locality" and so guided Yarrow to his conclusions. Baird fully expected that the flora and fauna of the new regions would quickly change, for he instructed Smithsonian collectors to obtain as full a complement of common species as possible for each region to record its onetime status.[60]

127

Nature
Writ Large
and Small:
Local, State,
and National
Natural
Histories,
1850–1875

Most poignantly, the inhabitant observers began to notice changes in their beloved environs. Howard Jones noted that his list of nesting birds differed considerably from that of Jared Kirtland, compiled some forty years previously, owing to the increase in cultivation. Kumlien personally observed the changes over three decades; in his article on the "rapid disappearance of Wisconsin wildflowers," he attributed the decline in the variety of flowers to farming: "a list of the plants of this vicinity, giving the plants of to-day, would be a comparatively meagre one and nearly useless, as their number is lessening every year, and a list of the plants of thirty years ago would perhaps have no other than a small historical value."[61]

The settlement that eventually converted the wilds initially entailed exploration of new areas and the planting of natural history disciplines. Harmon Atkins and Lewis Henry Morgan, originally from New York state, transferred their interests to new areas in the Midwest, to which they moved or traveled. Transplantation was not limited to the native-born, as Kumlien employed near a remote Wisconsin lake his skills honed in Linnaeus's hometown of Uppsala, and German-born George Engelmann investigated at his St. Louis home cacti collected over thousands of square miles. Collectors such as Annie Law and Mary Andrews used their varied travels as opportunities for further research, while Benjamin Wailes and P. R. Hoy capitalized on homegrown incidences to find their specimens. The inhabitants of new areas raised appreciation of their new home back east, whether in the Smithsonian Institution's publications of their findings or in books such as Increase Lapham's melding of Wisconsin boosterism and scientific cataloging, which were distributed nationwide.[62]

The Participants

Like the preceding generation, the artists of this period came from backgrounds so diverse that only their common adherence to conventions unified them. Some were naturalists who learned to draw for their own purposes, for example Sullivant, Morse (who learned to draw with both hands simultaneously, a technique he showed to great effect in lectures), and James Emerton (an entomologist who later carved a unique career of modeling squid for museums). Robert Ridgway as a boy drew and watercolored the birds he read about and observed. Other artists learned "on the job" for specific projects, as did Mary Peart for *Butterflies* and Mrs. Virginia Jones when she re-

128

THE BOOK
OF NATURE
Natural History
in the ·
United States

placed her late daughter as artist and lithographer for *Nests and Eggs of Birds of Ohio.* The expedition artists had vastly different backgrounds. The Kern brothers, Richard and Edward, who accompanied the Sitgreaves expedition and others, had previously illustrated medical botanical books. John Mix Stanley had no previous experience in scientific illustration before his botanical drawings for the Mexican boundary and Pacific Railroad surveys. Arthur Schott had studied natural history in his native Germany and so had been exposed to botanical illustration even before knowing his American mentor, John Torrey. John H. Richard's long career of drawing specimens foreshadowed the professionalization of natural history artists illustrating the federal government's agricultural and economic entomology publications at the turn of the century.[63]

The opportunity to observe fauna and flora in new locations brought prominence to those naturalists able to participate in published monographs and articles. This era marks the high point of the "nonprofessional" scientist, as those not associated with natural history as a career could still publish definitive books and articles and were leaders in their field. Like John Bachman, Moses Ashley Curtis miraculously fitted world-class study into the busy life of a parish clergyman. Lewis Henry Morgan combined his ethnological writings and beaver work with the career of successful businessman. William Starling Sullivant used his independent wealth to sponsor the artist Schrader and pay for plates in his works. Nelson and Howard Jones took time from their medical practices to observe nesting birds. Edwards somehow retained nominal presidency of a coal company through his adult life of copious note-taking, letter-writing, and egg- and larvae-raising.

Full-time naturalists like Gray and Louis Agassiz shared the same devotion to their subject, but the words of these authors more potently convey their absorption. Morgan joked that the beaver had "entrapped him," laughingly masking the energy and curiosity he poured into his efforts. The publication of the *American Beaver* proved a fitting climax to and sharing of his enthusiasm. Order and purpose in life, coupled with a sense of accomplishment, inform Edwards's writings. At the end of his entomological reminiscences, Edwards succinctly tells what studying United States butterflies meant to his life: "I have reached my eightieth birthday. My pursuits in natural history, first in ornithology, began when I was a sophomore at Williams College in 1839, and continued through 1846, next in lepidopterology, have contributed to make the years agreeable as they passed. I have been happy in my home life also. I should be willing to live over again, and I hope the next life, wherever it may be, may have pursuits as delightful as those which have occupied me here."[64]

129

Nature
Writ Large
and Small:
Local, State,
and National
Natural
Histories,
1850–1875

Historian Anne C. Rose has suggested that Victorians of the middle to upper class were arduously seeking "leisure that engaged the imagination" and desired a "radical . . . expansion of feeling and imagination cultivated within outwardly respectable lives." The traditional comforts afforded by religion and family life did not totally satisfy, but deliberate flouting of convention was not considered. Certainly these naturalists, following a profession yet devoting their energies to another activity, fit this proposed pattern. Others of their generation found interest and stimulation through different pursuits; it was the "delights" provided by natural history study that "entrapped" Morgan, Edwards, and Curtis throughout their lives.[65]

The efforts of local observers such as Kumlien and Atkins and collectors such as Law and Blaney gained the greatest prominence as well in this era. The national monographs still featured their discoveries and anecdotes, and the Smithsonian solicited these "intelligent" observers to send in samples. Although few in number and scattered throughout the nation, their efforts found home in collections and in catalogs. Those who sent in an odd fish for Agassiz's project on fishes helped the efforts, but the even smaller number who pursued the study truly shared the naturalists' perspective. Kumlien, for example, went beyond the stereotypical designation of sex roles in nesting when he noted the "naked" belly of the male Wilson pharalopes, indicating that it also incubated the eggs. Also contrary to expectations was mating females' aggressive fight for the male. Theodore Mead showed the determination and knowledge of his mentor Edwards in his successful capture of hundreds of butterflies on his expeditions and his bold asseverations concerning mating. Edwards's other correspondents had the savvy to distinguish the species in which he was particularly interested and possessed the patience, skill, and timing to obtain eggs of specific species outdoors and to raise them over food plants at home. (The efficiency of the railroad package services in delivering the live butterflies deserves mention, too.) Samuel Scudder, himself a butterfly expert and a contemporary of Edwards, knowingly salutes those who assisted Edwards in observing butterflies that would not reproduce indoors: "The agents had to remain on the inclement or wild spot long enough, often days, to secure eggs freely laid by an imprisoned female, whose moods are dependent on sunshine and a certain warmth." The few but vital Smithsonian correspondents cunningly captured and intrepidly handled potentially dangerous reptiles.[66]

130

THE BOOK
OF NATURE
Natural History
in the
United States

However, one may contend that land snail collectors such as Annie Law, who were able to gather the new species that their naturalist correspondents desired, demonstrated the most dedication to seemingly uninviting objects—monochromatic, small, and barely distinguishable from one another. The land snails' historian, William Green Binney, admitted, "in regard to colors, our snails are quite plain and exceedingly uniform . . . most of them being simply horn-colored." To discern differences between species indeed required skill. Edward S. Morse, whose early love was land snails, illuminated the subtler pleasures of collecting the subjects: "The soothing pleasures of lying down, dorsal region uppermost, in some secluded grove, and hunting for half a day among the decaying leaves . . . no frantic hops, skips, and jumps of the insect tribe, no terrible bites to dread . . . no giant stride or rapid speed to wonder at."[67]

These correspondents also sought intellectual stimulation and spiritual satisfaction in their new activity. The reward of time stolen from their occupations and families was a mastery of a complex subject. Mead and Law could converse with leaders of a discipline and make original contributions; Kumlien could share through ornithological literature his highly unexpected field observations about nesting and mating. The continuation of study enhanced such pleasures as enjoyment of natural beauty, and the sense of accomplishment increased with the effort expended.

The lack of financial reward underscores the dedication of the authors and the other participants. These ambitious monographs did not yield their creators profit because of the same publishing economics of previous times. The hand-coloring and painstaking lithography that the authors desired, coupled with the process of correcting proofs and coloring "long distance," as in the example of the Audubon works, made publishing a lengthy and awkward effort. (The Civil War intruded on various projects; for example, Elliot's publishing of his grouse folio slowed when the Bowen lithographer, C. F. Tholey, was drafted.) Edwards told Baird that the *Butterflies*, for which he obtained over two hundred subscriptions, "does not pay" because the coloring, lithography, payments to the artist, and fees to booksellers greatly lessened his profit from each volume sold. His preface, calling "works of this class . . . [not] remunerative in a pecuniary sense, [but] strictly a labor of love," reflects his experience. Elliot generally issued two hundred copies of his monographs and seemed satisfied with these sales; again, his independent wealth subsidized his time and probably started up the projects. Baird's Little, Brown & Company production of *North American Birds* did not adequately compensate him for his time, as this characteristically diplomatic man uncharacteristically complained to his colorist, Lavinia Bowen, "that [she] would be paid for her labors, as he would not." The price of five dollars per part did not pay for the total production costs of the *Nests and Eggs of Birds of Ohio*, and Dr. Nelson E. Jones, who underwrote the work, eventually lost over eleven thousand dollars, in large part because one-third of the copies remained unsold. He confessed to a publisher, "You may ask why such an arduous and expensive undertaking was consummated by one with moderate means—I know why, but no one else can ever understand."[68]

One lithography firm that executed the federal government survey plates complained that the time and labor spent in the finely detailed work so drastically minimized their profits that they took the job only to attract future "simpler work from other sources." The artists' fees probably did not remunerate their expertise and many hours of intensive labor. John H. Richard told Baird of his long days painstakingly executing the government expedition works (in his inimitable spelling, "I workt very hard and averick 11 haur par day"), and no amount of money could compensate Mary Peart for her efforts in raising and drawing larvae. For these coworkers as for the authors, the completion of a memorable project was the goal.[69]

Thus, many monographs remained fragile constructs of individuals' efforts. Even the state and federal government publications reflected distinct and identifiable contributions, from illustrations such as Arthur Schott's plucked desert plants to life history descriptions such as Robert Ridgway's "steel-blue" white-bellied swallow. Such publications as the *Zoology of New York* were attributed to authors; therefore, some intense criticism was directed to James DeKay, not to faceless government bureaucracies. Spencer Fullerton Baird likewise coordinated Smithsonian publications and expedition reports.

One of Baird's greatest achievements was his cultivation of contributors. He and other authors, such as William Henry Edwards and William Green Binney, successfully drew together the specimens and observations gathered by such geographically scattered correspondents as Annie Law, P. R. Hoy, and Theodore Mead. These individuals, like the leading naturalists, participated actively in natural history study for private, if not financial, gain. Because their contributions often reported the first observations of appearance and habits, they received thanks and publication.

The careers of naturalists have already suggested that reading and viewing transmit the natural history tradition. Baird, for example,

132

THE BOOK
OF NATURE
Natural History
in the
United States

had appreciated Audubon's oeuvre and applied its standards of high-quality illustration and thorough life histories to both the federal government works and his monograph on birds. William Henry Edwards, another Audubon devotee in his youth, applied these standards to butterflies. Because these illustrated monographs never enjoyed wide circulation, the number of people who learned natural history as Baird and Edwards did appears limited. However, forces outside the natural history disciplines between 1825 and 1875 were to enable dissemination and adaptation on a scale never before achieved.

4

Dissemination and Response:
Print Culture

*✎ I have seen some numbers of your work [the oc-
tavo edition of <u>Birds of America</u>] now publishing, and
admired them very much. . . . For my part I read the
descriptions on birds and the episodes in your "Ornithological Biography" with the same emotion of pleasure
as I used to read a favorite novel.*

Spencer Fullerton Baird in a letter to John James Audubon,
20 June 1840[1]

The illustrated natural history monographs, with their specialized language, high prices, and small press runs, suggest a limited "reading community." The tens of thousands of popular novels sold in the period from 1825 to 1875 dwarfed the number of the federal government surveys' editions, which, in turn, easily surpassed the quantity of subscription works distributed. The nature of the publishing industry and the cultural preoccupations of literate antebellum America, however, enabled the monographs to become known beyond their original audience of naturalists and subscribers. Textbooks, children's and sportsmen's literature, general periodicals, and other literary genres such as biblical natural histories and ladies' flower books were disseminating the orig-

134

THE BOOK
OF NATURE
Natural History
in the
United States

inal works of Wilson, Audubon, Godman, and the botanists by adapting or copying chunks of their material.[2]

This increased availability influenced not only individual pursuits but also generally held attitudes toward plants and animals. Textbooks and children's literature provided the initial stimulus to future devotees and deeply influenced those lives. Furthermore, writers and publishers so actively championed the increased knowledge of and empathy with animal and plant life that their wide range of readers became familiar with a host of species to an unprecedented degree. The discourse in general became so well known that even non-naturalists could employ it to express their encounters with the natural world.

Historians have long pointed to the popularity of natural history activities in these decades, but they have not attributed the growth in popularity to historical agents, among them the diffusion of print culture. The few who have studied the effect of natural history on discrete areas such as education and public lecturing perhaps have underestimated the central impact of the text upon the individual reader. Although classroom lectures and mentorships held crucial roles in imparting the disciplines, the text remained central because it held the exact terminology and species identifications that could be studied again and again to achieve concrete knowledge. Authors and editors basing their works on natural history usually edited out the Latinate terminology to make works "popular" but retained from the monographs some original principles, namely the life history or biographical sketch of species and the anecdotes of individual animals or plants in interesting situations. Thus, their readers still experienced natural history discourse.

Explanations of how and why audiences appreciated natural history literature in all its permutations have remained shadowy. This chapter attempts to capture these dissemination patterns and individual responses. Access to popular reading materials excluded nonliterate Americans and concentrated the dissemination in households receiving periodical literature and in areas where such materials were most available—namely in cities and in the Northeastern United States and along transportation routes in the expanding West. Stories of individuals from remote areas, however, dramatically illustrate the impact of natural history beyond the urban centers.[3]

The Natural Background

The nineteenth-century Anglo-American relationship to flora and fauna illuminates much of the motivation for using natural history discourse in popularly oriented literature. Cultural historians of Great Britain have emphasized the near-obsession with animals in the nineteenth century as manifested in activities and institutions such as pet breeding, hunting, and zoological gardens. Students of the United States also note the profound interest in plants and animals. Popular books on "animal intelligence" musing over the remarkable learning propensities of pets, as well as titles of artworks comparing canine to human attributes (Thomas Landseer's *Dignity and Impudence*, for example), consciously connected human and animal qualities. Even plants, whose insentient nature traditionally disassociated them from humanity, exhibited kindred traits to the Victorians, as in "tyrannous" tropical climbing vines that "displayed a spirit of restless selfishness, eager emulation, and craftiness" in a forest where "the motto of the majority is—as it is, and always has been, with human beings—'every one for himself, and the devil take the hindmost.' " Thus, writers and their audience found the self-image of human individuals and their communities, negative and positive, in the fauna and flora. In ways perhaps naive to today's readers more accustomed to scientific objectivity, Americans voiced conscious concerns over moral and social issues in their natural history adaptations.[4]

The process of industrialization in Britain and the United States fostered a feeling of nostalgia for "Nature" as urbanization and settlement irrevocably changed it. Humans, disturbed by rapid changes in all spheres of life, sought the same psychic comfort and refreshment in "Nature" hitherto provided by traditional religion, and they reiterated with a fresh vigor the natural theology argument that Nature was God's handiwork. In this transatlantic cultural shift, most often defined as "Romanticism" (which began in the eighteenth century but accelerated in the nineteenth), animals and plants, the quintessential creations of God, were especially deemed to hold spiritual meaning. Medieval bestiaries and the emblem books published until the early Republic had connected plants and animals to specific morals (e.g., the butterfly to brevity of life, the bee to industry), but as historian James Turner has noted, the newer attitude viewed animals as fuller role models, not one-dimensional synedoches. Their actions, especially those analogous to those of humans, like childrearing, could teach God's will or exemplify moral conduct.[5]

This growing sense of kinship fueled the animal humanitarian movement, coming to fruition in the establishment of the Royal Society for the Prevention of Cruelty to Animals in 1824 and its American counterpart in 1871. Individual Americans such as Bartram and Wilson had previously sought consideration for birds on the grounds of

136

THE BOOK
OF NATURE
Natural History
in the
United States

their perfection as God's creations, but the widespread intensification of interest in animals and plants did not begin in the United States until the 1830s. It became one constituent of an ethos originating in the Northeast in response to the "market revolution," the transformation of local economies to nationwide interests, which was instilling such businesslike values as competition and precision into everyday life. Absorption in "Nature" by any means, including interest in plants and animals, relieved the tensions of the new lifestyle in a fashion similar to the welcoming hearth and home created by the wife and mother refreshing the tired breadwinner (part of the market revolution response now labeled the "cult of domesticity").[6]

Natural history literature served the important function of conveying information about the all-important nonhuman world. Natural history monographs, particularly those of the ornithologists and botanists, held reservoirs of information about nature's increasingly cherished products. By adapting the monographs' species descriptions, zoological and botanical textbooks and their periodical excerpts in effect introduced potential moral exemplars and mirrors of human behavior. But the mechanisms for effective publication were required.

Patterns of Dissemination

The publishing industry was itself part of the "market revolution" undergoing significant expansion. The reduction of printing and paper costs through the antebellum decades, thanks to technological innovations such as advancements in the steam press, tempted would-be publishers to start a plethora of magazines and newspapers throughout the United States. Many of these sadly proved short-lived owing to insufficient capital. However, several book publishing firms in the major Northeastern cities, such as the Harper Brothers Company of New York, successfully exploited the technology and new transportation opportunities made possible by the railroad to build a nationwide market; thus, they became the dominant forces in textbooks, nonfiction, and novels written by Europeans and Americans.[7]

The increase in the number of titles published and copies produced held certain implications for natural history dissemination. The works of American authors proliferated in many genres, including those popularizing natural history; the lack of original material that had disturbed earlier cultural nationalists was disappearing.[8] More Americans were reading about native species as interpreted by "native" authors such as Alexander Wilson, John James Audubon, and Asa Gray and reinterpreted in textbooks and journals. The life histories, characteristic habitat, and personal observations—those "inhabitant" aspects of American natural history—were thus transmitted. However, the continuing practice of reprinting foreign books and placing these excerpts into periodicals ensured transatlantic influence—British in particular. European, African, and Asian species well represented in British natural history literature populated U.S. material as well. Americans thus could explore the wonders of fauna and flora of other continents through the medium of print.

Textbooks and Encyclopedias

Textbooks proved a most lucrative staple in antebellum publishers' lists. Ostensibly designed for "common and high schools" and college use, they were purchased by many individuals outside the classroom. Botany was incorporated slowly into the seminary, public school, and college curricula during this period while zoology was not. Therefore, many textbooks were purchased for individual use or as gifts for youngsters of both sexes. The educational designation assured prospective buyers, mostly parents and adult friends, of accuracy and morality. Serious learning was a paramount consideration for the gift givers: an elder cousin wrote on a flyleaf of a Peter Parley's botany textbook to its female recipient the potent Baconian dictum, "Knowledge is power." As one writer for Sabbath school books in the 1840s acknowledged, "There is hardly any a subject which more interests the mind of youth, than that of Natural History, particularly when it is illustrated by pictures." Adults knew that the subject held appeal apart from (or despite) its moral purity.[9]

The Harper Brothers Company adapted works originally published for the London Society for the Diffusion of Useful Knowledge by the British naturalist James Rennie, with the American editors claiming to "carefully revise and omit those portions least interesting." Editors transformed Rennie's works on insect and bird "architecture" into *The Natural History of Birds* and *The Natural History of Insects* by omitting a few technical drawings and inserting subheadings emphasizing the American citizen Alexander Wilson for the new American audience. These works were then packaged in Harper's famous "Family Library" series. Because the new technology of stereotyping that Harper helped pioneer in the United States created exact plates of pages, these editions (each with an initial print run of perhaps five thousand), first published around 1831, were published in the exact same format until 1859. Their many descriptions of the wondrous constructions of paper wasps and the intricate hanging

138

THE BOOK
OF NATURE
Natural History
in the
United States

nests of tropical birds might lead the reader to believe in animal reasoning power, despite Rennie's observation at the conclusion that only humans had the ability to learn. Rennie asserts the moral value of studying these "inferior" creatures, despite their reliance on mere instinct: "we may still take example from the diligence, the perserverance, and the cheerfulness which preside over the Architecture of Birds."[10]

The quasi-publishing concern belonging to Samuel Griswold Goodrich, known as "Peter Parley" to his legions of young readers, was responsible for the largest number of U.S. natural history textbook titles. Goodrich, who developed the persona of the wise and kindly Peter Parley to introduce his works to young people in the 1820s, established a stable of writers (at one time including Nathaniel Hawthorne) and a network of publishers, who aided him in producing over one hundred juvenile literature titles ranging from alphabets to geographies. He and his editors made in their natural history compilations "free use of the Library of Entertaining Knowledge [another American reissuing of the Society for the Diffusion of Useful Knowledge], the Family Library, Wilson's Ornithology, Godman's Natural History of North America, &c." The comprehensive "Parley" work on the animal kingdom, *The Naturalist's Library*, contained modified life histories of the mammals' physical appearances and habits plus wood engravings, including many exotics such as the tapir and ocelot, derived from imported texts on the highly popular English menageries at the Royal Zoological Society and the Tower of London. The inclusion of abundant adaptations of Wilson's and, in a few cases, Audubon's life histories resulted in extensive coverage of American birds. In contrast to this full coverage of birds and mammals, it showed only a few invertebrates, such as the horrifyingly fascinating "gigantic cockroach, a native and plague of the warm parts of Asia, Africa, and South America," perhaps because the available literature did not give much information, interesting or dull, on the majority of mollusk and insect species. Goodrich reused these passages in his *Parley's Book of Ornithology* and *Tales of Animals*. That *The Naturalist's Library* and these other works under various titles were issued in different locations and times from the 1830s to the 1860s was due to Goodrich's system of renting out his stereotyped plates to publishers.[11]

As its "swan song" to popularized zoology in 1859, the Parley concern recompiled *Illustrations of the Animal Kingdom* from more recent works including Audubon and Bachman's *Quadrupeds*, the entire Audubon *Birds*, John Cassin's *Birds of California*, the Pacific Railroad

reports, James DeKay's and Jared Kirtland's state works, and British
works such as the *English Cyclopedia of Natural History*. Latinate nomenclature makes an appearance, but as with the former Parley works, "a
just balance between the skeleton of system and classification" and
"an abundance of details, descriptions, [and] incidents" was sought.
Indeed, the life history adaptations of the species in question from
both British and American sources, sprinkled with firsthand accounts
from travelers and naturalists, dominate the texts. Goodrich claimed
in his autobiography that at least seven million copies of his works,
including fiction, history, and geography, were in circulation by
1856. If only a fraction of these works dealt with natural history, the
Parley productions would have made available the scientific works of
Wilson, Audubon, and others to a far larger audience than represented by the original subscribers to the scientific monographs.[12]

Other zoological textbooks were issued, ranging from the American editions of the English import John Bigland's *Natural Histories*
(published from 1828 to as late as 1865) to the more scholarly compilations of Samuel Ruschenberger, a Philadelphian physician and
member of the Academy of Natural Sciences, who translated contemporary French anatomists' natural history manuals for his series,
"First-Books of Natural History." The schoolteacher John Lee Comstock's 1853 *Readings in Zoology*, a successor to his *Natural History*, begins with a brief description of the races of man but devotes the bulk
of the book to birds because his son's "zeal and knowledge in that
department" contributed lengthy coverage. The younger Comstock
not only adapts physical descriptions, nests, songs, and geographical
range from Nuttall's *Manual* but also extensively quotes from the Audubon and Wilson texts to demonstrate their "poetical" and "nature-
loving" flavor in passages such as the "mutual love chatter" of the
meadowlark. The formidable husband-and-wife team of Sanborn and
Amy Abby Tenney wrote a series of zoological textbooks between
1865 and 1875 geared to readers of all ages. His *Manual* and *Elements
of Zoology* included thorough discussions of the invertebrates, thanks
to the works of recent malacologists and entomologists, but retained
"nearly complete catalogues" of birds and mammals thanks to Audubon, Bachman, and Wilson. Her works in the series *Pictures and Stories
of Animals for the Little Ones at Home*, for the youngest readers, introduced the study of marine life, widely popular a decade before in
England, to U.S. audiences.[13]

These zoological texts gave their readers a "taste" or a more comprehensive understanding of the subject but did not expect students
to participate actively in the discipline. Ornithology, ichthyology, and

140

THE BOOK
OF NATURE
Natural History
in the
United States

mammalogy at this time required extensive hunting, fishing, and trapping skills. Women (and the many men unfamiliar or unskilled in these pursuits) were thus immediately excluded. American women were expected to explore zoology mostly in the field of marine life, whose collecting required mainly patience and imperviousness to aquatic conditions, and in a few known examples, insects.[14]

Botany, by contrast, was eminently "do-able" for both sexes: Almira Lincoln Phelps, the prominent nineteenth-century textbook author, assured her readers that as her publications removed "obstacles which formerly impeded the progress of botanical information . . . among our own sex," "the temple of Flora" would finally open to women. The specimens for observing and collecting a variety of classes and families lay firmly anchored and easily plucked in the nearby outdoors. In addition, "of all sciences, perhaps no one is settled on a firmer foundation than that of Botany," in large part because of the extensive discoveries and publications of eighteenth- and nineteenth-century figures John and William Bartram, Nuttall, William Barton, and John Torrey. The textbooks fed beginners' interest by joining firsthand observations with identification and classification derived from the monographs. For example, Phelps's *Lectures on Botany*, first published in 1829, asked the reader to pick a specimen and then read about its parts from the text as an introduction to plant structure. The Linnaean, and, to a lesser extent, the "natural" system allowed the reader to insert actual specimens into a classification system based on observable parts.[15]

The initial success of pioneer science educator Amos Eaton's *Botany* (he boasted that over 2,500 copies had been distributed to students by 1825) stimulated a host of educators to share the market. "Parley" issued the same botanical text derived from Eaton and the eminent botanist John Torrey in different guises for three decades, and schoolteachers John Lee Comstock and Alphonso Wood issued their introductions to botany. John Darby and the New York firm of Derby & Company issued a manual in 1854 developed for Southern schools and colleges, because "an agricultural people" were supposedly interested in botany. James Rennie's *Alphabet of Botany* was a popular import.[16]

Most concentrated on straightforward presentation of physiology followed by extensive presentation of all the plant kingdom's classifications and the most important genera therein. Few anecdotes were allowed, since so much taxonomy filled the texts; nonetheless, habitats and the most common U.S. species were indicated. Most carefully adapted the authoritative botanists; in fact, the preeminent U.S. bota-

nist, Asa Gray, wrote *A Manual of the Botany of the Northern United States,*
Elements of Botany, and *How the Plants Grow,* which enjoyed many editions and reissues. Gray privately referred to them as "hornbooks"
(primers) to professional colleagues and implied that he wrote them
for income, but his uncompromising emphasis on taxonomy and
Isaac Sprague's abundant wood engravings ensured that readers were
receiving the most accurate information available. The number of
botanical textbooks eventually sold in the antebellum United States
was enormous compared to those first runs of Eaton. Gray's *Botanies*
ran to over ten editions, often five printings each. Phelps's sales
topped 375,000 copies by the thirty-ninth printing of her *Familiar
Lectures.* Wood's *Class Book of Botany,* first published in 1845, sold between 800,000 and 1,000,000 copies in its various editions.[17]

Encyclopedias proved another vehicle from which to learn natural history. Following in the British tradition, American editors sponsored copious entries by leading naturalists such as Thomas Say and
John Godman, who contributed substantial articles to the *Encyclopedia
Americana.* Amos Eaton told his readers of *Textbook on Zoology* (1822)
to supplement his genera descriptions by consulting *Rees Cyclopedia*
for species, because it was "the only English work, to be found in our
towns and villages." Asa Gray's first exposure to botany came not
from a teacher or textbook but from an article printed in the *Edinburgh Encyclopedia,* edited by David Brewster. As a youth on a farm in
upstate New York, he borrowed the volume from a young men's society's circulating library and read the densely packed 232 pages on
the classification, physiology, and history of botany. William and Eliza
Sullivant first began their studies in botany and conchology with
those entries in the American edition of the *British Encyclopedia* (1819,
1821), which became "well worn," according to their biographer.
The young Spencer Fullerton Baird studied "the Comparative Anatomy of Birds & Quadrupeds . . . using the article on Birds in Rees
Cyclopedia" but eventually asked his mentor Audubon for another
reference after he had exhausted the resource with characteristic
thoroughness: "not being able to make head or tail out of the descriptions of the muscles &c., I threw it aside & wrote one myself."[18]

Periodical Literature

Natural history articles proved a godsend to editors, who, searching for material to fill columns, "cut and pasted" from other books
and magazines without fear of copyright infringement. (One newspaper subscriber in 1839 complained that "the country is flooded with
periodicals, the one containing in a great measure the ditto of the

142

THE BOOK
OF NATURE
Natural History
in the
United States

other.") They were gearing publications to a rapidly diverging audience, yet natural history extracts appealed to all or at least offended few. The connection to God made natural history eminently suitable for the heavily moralistic children's, family, and "ladies' " fare. Because the God invoked fervently yet hazily in many pieces was "Nature's God," no sect could take offense in these decades of religious feuding and virulent anti-Catholicism. The outdoor flavor suited the influential agricultural and sporting presses. In the antebellum era of rising sectional tensions, natural history avoided divisions between North and South by deflecting attention to a noncontroversial subject. Items on foreign or newly discovered animals in the American West appealed as "exotics" to both readerships, and thanks to plant and fauna distribution and migration, many of the common species were shared.[19]

Natural history anecdotes, in particular, offered outré tidbits morally acceptable to the public. Their brevity and human interest caused them to be widely reprinted in a variety of vehicles, while their firsthand nature experiences lent scientific veracity many appreciated. For example, Francis Woodworth told an anecdote gleaned from a friend, "who is a very close observer of the lower animals," of a drake easing off a tin strip stuck on a goose's neck. He repeated this story frequently in his magazines and books, and other children's magazines picked it up. The *Lowell Courier* ran an article of two gentlemen watching a weasel catch a rabbit in the deep snow: the weasel surprised the rabbit by burrowing beneath it (*Forester's Children's Magazine* included this anecdote in its 1842 issue). The Philadelphia *Saturday Courier* in 1843 reprinted a letter to the Baltimore *Republican* about a colony of swallows in Cecil County, Maryland, more specifically, "Mr. Stump's farm." The unprecedented and "remarkable occurrence" of at least thirty-seven pairs building their nests in a straight row under the barn's eaves so stirred the letter writer that he made the effort to read Audubon to learn more about the new birds. Animal intelligence and emotions were of so much interest that readers contributed their own anecdotes. "J. B." wrote to the *New England Farmer* how he observed a bluebird pair feed an orphaned blackbird: "[T]hat beautiful passage of scripture flashed upon my mind—'Are not five sparrows sold for two farthings? and not one of them is forgotten before God?' " The correspondent assured the newspaper editor that "many of my neighbors could testify to the above facts," as they had witnessed the interesting phenomenon.[20]

Editors' fascination with the "differentness" or "otherness" of animals explains many choices of articles. *Ballou's Drawing Room Com-*

panion in the 1850s often ran pieces, adapted from British natural history texts, featuring the "rare and Curious birds" and "an Interesting Illustrated Series of Various Specimens of Fishes." Its subscribers were treated to many paragraphs on the habits and appearances of the "wingless bird" of New Zealand, "known to aborigines as Kiwi-Kiwi," and the "singular looking Iguana . . . perfectly amiable in their deportment, unless crowded upon." Harland Coultas, a teacher who adapted his popular natural history books for *Godey's Lady's Book* in the 1850s, contributes "Curiosities of the Vegetable Kingdom" in which he explains how "the leaves of this plant [the moving plant of British India] are in constant motion by night and by day."[21]

Such articles assumed that these creatures would satisfy their readers' desire for experiences different from those of everyday existence. This attraction to the curious in natural history echoes earlier publications such as the English eighteenth-century *Gentleman's Magazine*, which ran excerpts and adapted illustrations from Catesby's *History of Carolina* to complement its articles on other foreign curiosities. The need to enliven burgeoning magazines and "penny" and "mammoth monthly" newspapers through the 1830s to 1850s reinvigorated the interest. In general, the reading public was increasingly exposed to the "sensational"—crime stories, accidents, mysterious murders—through the expansion of the print medium. Tales like that of the moving plant and the imprisoned goose, if less dramatic, display the same appeal.[22]

Animal anecdotes also reveal qualities and actions akin to those of humans. This similarity, as well as the "otherness," intrigued editors and readers. The letter about the bluebirds and crow, for example, was entitled "Benevolence in Birds." A Milwaukee newspaper quoted Wilson's anecdotes about a crow's cleverness. The titles of pieces in the New York newspaper *Albion*, excerpted from a copy of the *Ornithology Biography* provided by Audubon himself, show the interest in resemblance to human nature: "The Sagacity of the Canada Goose" and "The Goose—Its Loves and Gallantry." *Ballou's Pictorial Magazine* ran illustrations and texts adapted from the Audubon blue jay and brown thrasher plates: of the image of a black snake attacking the latter's nest, the editors praised the "bird-tragedy, as it were, that thrills the heart of the spectator, like some sad event in human life." The amazing bird architecture of the bottle titmouse was featured in the *Child's Cabinet* as evidence of bird intelligence and was one of many "hanging bird" nests in a *Ballou's* feature. (Not all animals' mental capacity was deemed outstanding, as in a *Godey's* report that the musk deer's "full dark eye gives it the appearance of a degree of

144

THE BOOK
OF NATURE
Natural History
in the
United States

intelligence which it does not possess, for the greater apart of its time is passed in eating, drinking, and sleeping.") Harland Coultas in *Godey's* even emphasizes the "analogies between the Animal and vegetable kingdom" by describing the zoophytes (coral and sponges) as demonstrating "many striking indications of a vegetable nature": "There is abundant reason for believing that the differences among the organic productions of nature are not so great as once supposed."[23]

Descriptions of life cycles granted autonomy and respect to those species, for they implied that plants and animals led interesting existences beyond human influence. For example, *Graham's Magazine* "for the Gentlemen and Ladies" and *Godey's Lady's Book* in the 1840s and 1850s offered several series of life histories abbreviated from natural history texts. Graham's multiple installments of "Game Birds of America" and "Wild Birds of America" gave descriptions mostly from Thomas Nuttall's *Manual of Ornithology* with ample quotes from the birds' "best biographer, Wilson." Included was Wilson's moving observation that female rails sat on their nests to the death as northeast tempests flooded the marshes. The "Wild Birds" articles include common species such as the bluebird, lesser-known birds such as the black-throated green warbler, and, in a geographical stretch, the little auk who frequents Greenland. *Godey's* attempted to enlarge its readership's knowledge of everyday articles in "Shells for the Ladies, and Where They Come From," "Perfumes for the Ladies, and Where They Come From," and "Furs for the Ladies, and Where They Come From." The animals that once inhabited the shells now residing on mantelpieces and their original habitat are described: "[W]hen suddenly alarmed or threatened with instant danger, the mollusk [the ventricose harp] forcibly draws itself up into the recesses of its shell, dismembering itself of the posterior part of the foot." Such "biographies" describing the species from birth until death, although shortened from original sources, gave the subjects an identity and importance beyond their usefulness to humans.[24]

Editors sought to give their consumers an "expansion of feeling" through reading about different creatures and sharing empathic moments with these new characters similar to that derived from sentimental novels of the period. The emphasis on family life and romantic love is unmistakable and, given the overwhelming presence of these themes in the other features of the magazines, clearly intentional on part of the editors. The era of the 1830s to 1850s marked the rise of sentimental novels, in which authors constructed their tales to draw out sympathetic emotions from their audiences. The similar popularity of magazine short stories exalting the sentimental makes more apparent to the modern reader this aspect of the appeal of natural history articles.

An "expansion of empirical knowledge" also poured through these accurate abridgments. Editors and authors, pressured by deadlines, were content to quote chunks without extensive rewriting; thus, much factual and detailed knowledge of habitat, distribution, and physical appearance of a variety of species was offered. The articles imply that general readers held a different type of knowledge of the wildlife around them than that expressed in natural history (just as Wilson had thought decades earlier), and that they therefore required these simplified life histories. Naturalists continued to complain that most Americans, rural and urban, knew a few common species by widely varying local names and rarely knew their habits. A future ornithologist remembered that when he was growing up in the 1850s, "the man who knew the names of more than a half dozen common birds was rare indeed." Audubon, on a *Quadrupeds* promotion tour, joked that the "Great Folks" in Washington "call[ed] the Rats Squirrels, . . . Marmots, poor things, are regularly called Beavers or Musk Rats!" Such an audience would have found in the definitive names and detailed descriptions of activities such as nesting and migration provided by natural history texts an experience apart from their vernacular and popular attitudes.[25]

Among the available scientific works, those of the "field naturalists," who studied species in their natural environs, were most adaptable to popularization. English observers such as William John Broderip, author of *Leaves of the Notebook of a Naturalist*, stressed their own fascination with the subject by recounting their personal outdoor experiences. British and American reviewers acknowledged that such firsthand accounts "written by persons who describe what themselves have witnessed [better] appealed to the feelings, and rendered [natural history] attractive by the drapery of the imagination." For a work to reach a popular audience, "the reader must enter into his feelings and sympathize with them, or will write in vain for the majority." These narrative aspects in their writing coupled with their availability made Wilson and Audubon (most often via Thomas Nuttall's *Manual*) the most-quoted U.S. naturalists. Their copious works offered many incidents, and the style of their life histories often exuded "poetry" and "enthusiasm," two qualities highly prized in this age of sentimentality. Both frequently implored the reader to join in their emotions, as Audubon often implored his "gentle reader" to "imagine this . . ." Their wealth of visual and aural detail similarly engaged

146

THE BOOK
OF NATURE
Natural History
in the
United States

the reader's senses. Learned compilers who concentrated on synonymy and the fine points of nomenclature, in contrast, were admired but not excerpted.[26]

A distinctive image of the American field naturalist arose, owing to the frequent invocation of Audubon and Wilson. Audubon's original English admirers viewed the flowing-haired naturalist, with his unusual accent and "eagle mien," as an "American Woodsman," a sobriquet Audubon adopted willingly to gain attention. He wrote in the *Ornithological Biography*, for example, that the "life of a naturalist" after a day of hunting and drawing specimens ends in contented sleep by the campfire. Newspapers in England and the United States reprinted such scenes and ran the letters and journals from his expeditions to Labrador and the Missouri River much as they promoted the exploits of explorers like John Frémont. One admirer in the *American Turf Register* praised Wilson as "the eloquent pioneer poet of the woods, swamps, bays and fields," while the appearance of a disheveled Audubon shocking the clientele at a fashionable Niagara hotel provoked an article widely reprinted through the American press. Even more than to the frontiersmen Boone and Crockett, to Wilson and Audubon were attributed the moral qualities associated with nature, like simplicity, freshness, and holiness. Art critic Henry Tuckerman claimed in an article about Audubon that naturalists held "a most attractive combination of the child, the hero, and the poet—with, too often, a shade of the martyr." They were not the stereotypical rough "mountain men" of the popular press but rather the "interpreters of Nature's God," to quote one Audubon admirer. The article on the Niagara hotel incident vindicated the naturalist, "whose fame will be growing brighter, when the fashionables who laughed at him, shall have perished and been forgotten."[27]

Painters and printmakers created a corresponding visual iconography for the American field naturalist. Portraitists of Wilson and Audubon adapted the well-established conventional image of the English gentleman-sportsman holding a gun to suit the American frontiersmen. Such characteristics of Audubon as his long hair, fur coat, and open hunter's shirt, so different from conventional upper-class garb, signified further his outdoor lifestyle. For instance, *Gleason's Pictorial Magazine* adapted an English print of a well-manicured Audubon thus attired clasping his gun to his breast in a rhetorical rather than practical fashion (Illustration 4.1). The surrounding forest and stream and crowd of animals and birds visually associate him with his subjects. The outward gaze and wide brow indicate both spirituality and intelligence.[28]

Aside from the "enrapt scientist" anecdotes associated with Nuttall, no other American naturalists received as much attention or definition as Audubon and Wilson. The works of other "field naturalists" such as Thomas Say or the government naturalists were too limited, scattered, or dull to garner frequent newspaper extracts. The focus on taxonomy by top botanists such as Gray and the malacologists similarly decreased their attractiveness. *Gleason's* editors, however, bravely made one attempt to heroize a botanist. "Professor Shelton, the California Botanist," by increasing knowledge of that area's natural productions, they proclaimed, "was doing more to develop the real resources of California . . . than every quartz crushing machine, pickaxe and long tom, from Klamath to San Diego." The accompanying wood engraving shows a man in botanical collecting attire holding a freshly gathered specimen (Illustration 4.2). The vasculum (round tin box) and pickax take the place of the ornithologist's gun. One wonders whether this pleasant, dignified personage held as much interest for the audience as bolder frontiersmen types.[29]

Literary Genres

Literary genres popular in antebellum America adapted natural history, especially the life history format, for their specific purposes.

Illus. 4.1. "Portrait of Audubon, the Naturalist." Wood engraving by anonymous artist, for *Gleason's Pictorial Magazine* 3 (1852): 196. (Courtesy of the Winterthur Library, Printed Book and Periodical Collection.)

148

THE BOOK
OF NATURE
Natural History
in the
United States

Religious and children's literature, for example, chose stories and attributes to impart their moral imperatives. Sporting literature and ladies' flower books spread their renditions to different audiences. Readers of these works, in turn, were exposed to natural history facts and attitudes, albeit secondhand.

The now-forgotten genre of Biblical natural history represents such a usage. Imported from Europe and popular until the 1860s, natural histories of the Bible described the creatures mentioned in the sacred works with information from natural history literature so that the believer could realize more fully the Scripture. In the United States, this usage of natural history had the added bonus of elevating the subject. The Harvard theologian Thaddeus Mason Harris revised his popular *Natural History of the Bible*, first published in 1793, in an 1820 edition with illustrations. It lists alphabetically and gives brief

Illus. 4.2. "Mr. Shelton, the California Botanist." Wood engraving by anonymous artist, for *Gleason's Pictorial Magazine* 7 (1854): 165. (Courtesy of the Winterthur Library, Printed Book and Periodical Collection.)

information on the "beasts, birds, fishes, insects, reptiles, trees, plants, metals, stones, etc." in hopes that these references "serve to clear up many obscure passages, solve many difficulties, correct wrongs, . . . and open new beauties in that sacred treasure."[30]

Meanwhile, a home-grown effort was developing in the remote town of Blue Hill, Maine. Calvinist preacher Jonathan Fisher (1768–1847), who painted still lifes and peddled his own stories and poetry on his journeys through his parish, found time to write and learn wood engraving for his own biblical natural history, *Scripture Animals*. He modeled his engravings and abbreviated life histories of exotic species on old texts and watercolors that he had copied from the works of George Edwards and Thomas Bewick and European travel books from his student days at Harvard some thirty years before. He also spiced descriptions of American species such as the field mouse, far removed from the Holy Land, with his own observations. As preface to his poem warning humankind not to store riches on earth because the moth could "waste the store," he provided details of how a caterpillar he had placed in a box and fed daily with fresh leaves metamorphosed. After Boston publishers rebuffed his efforts, a Portland, Maine, firm issued his book in 1834. Fisher modestly "laid no claim to elegance" in his wood engraving (rightly so, perhaps), but his early love of natural history informs every sentence and engraved line.[31]

Later productions such as Henry Harbaugh's 1854 *Birds of the Bible*, more elegant in illustration and binding, expanded the theme that nature actively taught humanity. Harbaugh reasoned that because natural objects were created before the Bible and were "already given those characteristics which made it possible and proper to make them representative of divine wisdom to man," the Creator used animals to "illustrate and communicate" the Book of Revelation. Therefore, the spiritual reference to "dove eyes" conveys the love of the mother church because doves express tender love in their actions. The stork in biblical references was meant to signify mercy and kindness, "and the naturalist assures us that the stork bestows special care upon the education of its young." The cormorant was a negative role model, "a filthy predator": "As we would say, be not like Judas, but like John; so we say, be like the stork, but not like the cormorant." These lower animals and plants expressed more purely God's intentions because they had not fallen from grace. Instinct, according to Harbaugh, may more clearly "lead them right" than human intelligence. Destruction of such heavenly bearers could not be taken lightly. After reprinting the passage from Audubon in which

150

THE BOOK
OF NATURE
Natural History
in the
United States

he kills the mother kite that saved its young, Harbaugh chides him, for "we know at once how the matter strikes us, when we conceive the idea of beings above us coming down [and putting us to death] to inquire into the human constitution and economy."[32]

The belief that natural objects signified moral statements found its fullest expression in another popular antebellum genre, the ladies' flower book. These books usually comprise sections, each devoted to one flower, describing that species' physical appearance, the history and allusions pertaining to it, and poems dedicated to the flower. A list of the moral attributes given to specific flora, which constitute the "language of flowers," usually concludes the volumes. Such books match hundreds of traits to flora, such as deceitful beauty to the tulip, matrimony to the American linden, and patriotism to the American elm. Many traits were inherited from the European "language," but new American plants gave opportunity for authors to exercise considerable craft. For example, Sarah Hale, editor of *Godey's* and author of *Flora's Interpreter,* attributed "benevolence" to the Carolina allspice because Thomas Nuttall's description in *Genera of the Plants of the United States* stated that extra flowers bloomed after the terminal buds were cut at the end of the season.[33]

Other concessions to the American audience occurred in this imported genre. Authorship and material often favored home-grown products. Cultural nationalism received a boost from the frequent inclusion of "native poets" rather than the "hackneyed extracts from the standard writers of Great Britain," as in Lucy Hooper's work, *Lady's Book of Flowers and Poetry.* Emma Embury's *American Wildflowers* features original poetry dedicated to its twenty flowers. Higher values, according to one author, also belonged to America: "But in our republican country, where aristocratic distinctions among men are discarded, we will not attempt to introduce orders of nobility among the plants." In the list she appends to *Familiar Lectures,* Almira Lincoln Phelps explains the modifications of some of those suspect European traits: "[A]lterations have been made, in order to introduce sentiments of a more refined and elevated character, than such as relate to mere personal attractions." For example, the crown-imperial, a large flower of unusual beauty but giving off a "fetid odor," was linked to "Power without goodness."[34]

Because the "language" varied considerably from book to book, would-be lovers must have had a difficult time piecing together or deciphering a bouquet. Most likely, readers savored the more generalized associations of beauty and youth, for as Hooper said in the *Lady's Book,* "wild flowers of the fields . . . bring back to us a thousand bright recollections of sunny and rambling hours, when a fair un-clouded future lay before us." The format apparently was deeply satis-fying to its audience, if judged by the hundreds of titles sold. As convincing evidence of its popularity was its great similarity to the albums that also combined flower poetry and visual imagery crafted by female readers. Although often trite to us, the poetry pleased at least some readers, who marked "optimo!" in the margins to voice approval.[35]

Authors Hale, Hooper, and others sought to introduce botany to their female readers, although in limited doses. Hale briefly places the plants in class and order and describes their physical attributes. Hooper adds "a botanical introduction" listing the Linnaean classi-fications to her book of poetry. The title of John B. Newman's *Boudoir Botany* (1848) suggests its adaptation of others' botanies not only to an audience but even to a place. Authors frequently included Latin-ate nomenclature and suggested to interested readers that botanical manuals offered an admirable though different way of learning about favorite flowers. These ladies' flower books aimed to enrich women's empirical knowledge by first appealing to the sentiments.[36]

An aggressively male audience was reading life histories through quite a different genre. Writers and editors in the 1830s and 1840s were busily refashioning hunting and fishing as suitable recreations for gentlemen. Editors of the *American Turf Register* and the *Spirit of the Times*, and authors like Elisha Lewis and William Henry Herbert (best known to his followers as "Frank Forester"), were attempting to instill a sportsman's code stressing challenge and fairness into ac-tivities hitherto deemed utilitarian. As opposed to "pot hunters" (nameless and faceless beings who shot every creature, including small songbirds and "varmints" like frogs and weasels, "from the love of destruction"), their target audience consisted of urban "over-worked doctors, merchants, and artisans" and country squires who pursued game for the intellectual stimulation and contact with the outdoors. The sportswriters urged readers to avoid the "mass destruc-tion" methods like building pens and hunting game birds in the spring and instead to rely on shooting and angling accuracy.[37]

Editors included hunting scenes from naturalists in their maga-zine columns, favorites being Audubon's turkey and deer hunt de-scriptions. Lewis and Herbert molded their books on the life history format as they gave physical descriptions, range, breeding, and favor-ite haunts of each of the game species. Both admitted their reliance on Nuttall, Wilson, and Audubon for their game-bird texts, and in his *Game Fishes*, first published in 1849, Herbert proudly acknowledged

152

THE BOOK
OF NATURE
Natural History
in the
United States

personal favors from Louis Agassiz and the works of eminent British naturalists Sir John Richardson and Thomas Yarrell. These authors deliberately included extensive detail on appearance and characteristics in their virtual life histories. Lewis in his *Hints to Sportsmen* explains the rationale:

> We have already devoted much space, perhaps too much, to the natural history, habits, and peculiarities of the partridge, but, we trust, not without some benefit to our readers, as no one can expect to become an accomplished sportsman without studying very closely the individual characteristics of every species of game that he pursues. For it is by this knowledge, either gained through great labor in the field, or acquired from the writing and associations of those who have devoted the leisure of years to this healthful recreation, that one shooter is seen to excel another in the style of hunting and bagging his game.[38]

The discussions of nomenclature, European analogue species, and migration did not directly influence hunting practices, but they broadened the knowledge of this new type of "intelligent and scientific" sportsman. Herbert hoped that sportsmen would reciprocate the favor and themselves contribute to natural history. He delighted in proposing new fish species and in criticizing errant naturalists (for example, he called DeKay's ichthyological plates "atrociously executed") and ambitiously suggested that anglers take regular memoranda of "the form of the gill-covers; the number of rays in each of the several fins; and especially the form of the caudal fin-tail" of each fish caught so the "the many unknown and unsuspected" species could be discovered.[39]

Children's literature had different objectives and methods. Several formats disseminated natural history to younger audiences. Brief descriptions of animals and birds matched with pictures had been an important staple of children's literature since its development as a separate market in eighteenth-century England. Thomas Bewick originally intended his 1792 *History of Quadrupeds* only for the young, so its popularity with adults surprised him. Early republic printer-publishers relied on miniature "books of birds" and "books of insects" to appeal to beginning readers. These books, derived from the editions of the most readily available natural history book, Oliver Goldsmith's *History of Animated Nature,* had simple woodcuts and short texts sprinkled with curious facts, such as "In the year 1650, a cloud

of locusts entered Russia, in three different places." Printers outside
the major cities, such as the Phinney family of Cooperstown, New
York, continued until the 1850s to publish these chapbooks with stereotyped illustrations often derived from Thomas Bewick. Larger
publishers continued to issue books such as the 1855 *Birds of the Woodland* with short texts derived from the newer naturalists (for example,
Audubon's observation of the whippoorwill using its large mouth to
move its eggs was inserted).[40]

These picture books were straightforward and factual, with only
a few moral nods to nature as God's creation. However, writers associated with the booming evangelical Protestant reform movement
sought to speak directly to the heart of the young reader and persuade him or her to do good out of inward conviction rather than
from fear of outer corporal punishment. Conversations between
adults and children—and in the case of animal books, animals and
children—became a favorite compositional device in their books to
impart concrete knowledge and sway the emotions. For example,
Harper Brothers' character "Uncle Philip," who starred in his own
series of books, told the young men in his village about "animals that
know how to work with tools like a man," in particular birds "who
weave their nests and bees who make hives." (Wilson, described as "a
gentleman who has written a great deal concerning the birds of our
country," is the author most frequently cited.) Admiration of "these
tools to work with for its comfort, as good and perfect as any that man
can make," should inspire appreciation for God's craft and humility
("we should not be so proud of what we know"). Reprinted widely in
the United States, Emily Taylor's hugely popular British work *The Boy
and the Birds* translated into dialogue natural history facts such as the
metamorphosis of the Cecropia moth and the shape of a woodpecker's tarsus. The woodpecker, in fact ("I myself, the green woodpecker, or *picus viridus*"), quotes Alexander Wilson to tell of his own
usefulness in destroying forest insects.[41]

The children's writers most intently wished to curb boys' cruelty
to animals—teasing animals, taking eggs from nests, and shooting
birds—because such cruelty was deemed to foreshadow a heartless
adult nature. Reading about the activities and family life of a species
could induce empathy for individual animals. For example, in one
book "Peter Parley" ends the description of the mockingbird, accompanied by a wood engraving of Audubon's birds under attack from a
snake, with such an admonishment: "It is impossible not to feel a
sympathy with these poor birds, as we see them in the picture but we
suppose they would feel quite as badly to have their nest destroyed,

154

THE BOOK
OF NATURE
Natural History
in the
United States

or their eggs stolen, or their young ones carried off, by a monster called Ben, Bill, or John, as by a monster named Rattlesnake."[42]

Common Themes

Because the popularized natural history was extracted from a discrete corpus (the works of British field naturalists and American ornithologists and botanical manuals) and was liberally requoted, coherent themes run through the decades 1830 to 1870. Editors and writers steadily exposed different segments of the American reading public to distinctive elements of natural history according to their publishing agendas. They usually excluded the technicalities related to Latinate nomenclature, as their audience was thought not to have immediate interest in synonymy or minute distinctions between species; however, binomial nomenclature accompanied some common names, as was the case in the flower books. In contrast, anecdotes of individual specimens engaged in unusual activities were frequently reprinted from natural history texts because their narrative and curious nature were assumed to appeal to most readers. The true-story aspect satisfied sensational cravings, while this undoubted veracity made them purer to introduce to children than fiction, as Comstock explained in a book of animal anecdotes: "The study of Nature . . . is always the study of truth." In addition, the concrete specificity of examples made principles and themes more memorable: for example, in the first lesson of *Familiar Lectures*, Phelps instructs her readers to place a flower in one of their hands.[43]

Certain parts of the nonhuman creation received greater attention than others because of a general authorial climate that favored "associations" of objects with memories and emotions. Flowers, connected with many emblematical qualities and suffused with childhood and romantic remembrances, and birds, who exhibited so many parental and romantic qualities that the writer Thomas Bangs Thorpe called them "almost Christianized," outweighed the invertebrates and mammals (excluding domestic animals) in amount of "press." (One sports author admitted that, try as he might, he could not associate much maternal feeling to a fish laying eggs as compared to a quail decoying a hunter from her brood.) The simultaneous availability of popular botanical literatures (as in the case of Harland Coultas's contributions to *Godey's*) and bounteous ornithological texts enabled these adaptations.[44]

Most textbooks and periodicals so regularly blended foreign with U.S. species that Americans were exposed to as much information on

exotic shells, iguanas, and leopards as they were to descriptions of
American species. Thanks to *Godey's* series "Furs for the Ladies,"
readers learned probably for the first time the attitudes, distribution,
and food of the nutria and the European lynx. The practice of pre-
senting species in classification groups (e.g., all woodpeckers together) rather than by geographical area also diminished physical
distance. "Peter Parley" was not exaggerating when he claimed that
schoolchildren knew the exotic species through their books so well
that they could identify the animals on a first visit to a menagerie. A
Philadelphia gentleman in the 1850s chose to regale schoolchildren
with anecdotes about the elephant's memory and tiger's ferocity
which he derived from adaptations of Oliver Goldsmith's still-popular
History of the Earth and Animated Nature. Thus, popularization widened
the American naturalists' concentration on U.S. species.[45]

The literature associated with flowers, however, began to stress
native "wild flowers" as opposed to imported or domesticated varieties. The authors feared popularity of exotic plants threatened appreciation of the species located nearby but often overlooked: "How
few among us care to notice the wild flowers of our country; not that
they want beauty, for did this come from a foreign clime no expense
would be spared to cherish and preserve it, but being so close to our
homes it is overlooked in consequence of that very nearness that
should make us love it more." Just as writers of flower books lauded
the poetic efforts of American authors, they promoted "indigenous"
species such as the Carolina allspice and, in the case of Emma Embury's *American Wildflowers in Their Native Haunts,* devoted works to
U.S. flowers and scenery. In the decades ahead, garden book authors
would advise women to preserve the "old-fashioned" native species
by planting them in their gardens. National pride evolved into a personal concern.[46]

The analogy of nonhuman to human life cycles and a wealth of
empirical detail, particularly on "family" life, supported the humanitarian argument that animals could suffer as humans could and so
should not be mistreated. The adaptations of bird life history,
through their emphasis of the "natural" life of the bird outdoors,
also advocated preservation. Children's authors extended warnings
against hurting birds to include caging them, when they reasoned
that birds were most "at home" and led lives worthy of respect in
their native element. *Spring Lessons: or Easy Lessons for Young Children*
expressed the thought, "Do you not think they must look very pretty,
flying from tree to tree in the woods, from whence it is a pity to take
and confine them in a cage?"[47]

156

THE BOOK
OF NATURE
Natural History
in the
United States

One significant consequence of appreciating life in the original habitat was the gradual decline of keeping wild animals in the home. No longer would a Mrs. Messer engage in the risky undertaking of raising fledgling Baltimore orioles. These life histories' persistent argument of the smaller birds' economic importance as insect destroyers coupled with the continual analogies to human life made shooting and eating nongame birds increasingly unacceptable. In the 1850s one could find finches and orioles in the markets, but by 1869 a hunting manual advised its adherents to spare the catbird, bluebirds, and other songbirds, as "they do no harm" and "eat insects." The author did mention that the mockingbird's "flesh is sometimes eaten, but it is not very good."[48]

The theme of change and loss is as commonplace and widespread in popular literature as in naturalists' monographs. For example, an English reviewer in a general literature magazine cites the importance of Audubon's *Birds* as a record by which future generations could "institute a comparison for the purpose of ascertaining what changes civilization produces in the Fauna of that great continent." "Frank Forester" warns that more game laws were needed: "For profit, for pleasure, for mere recklessness and the love of useless slaughter, the work of extermination is going on eastward, and westward, from the salmon rivers and trout streams of New Brunswick and Nova Scotia, to the prairies and plains at the foot of the Rocky Mountains." A poem dedicated to Audubon in a Philadelphia newspaper in 1843 sighs, "How soon must perish every trace / of them [animals], but that which skill, / And taste, and feeling, yet may give."[49]

Arthur's Home Magazine, intended for the parlors of ladies and gentlemen, seems an unlikely forum in which to discuss the destruction of wildlife, but Charles Wilkins Webber's article on the pinneated grouse leads off the November 1852 issue. He laments the species' disappearance from the Northeast and notes its former large numbers in Kentucky, where his parents remembered them as being so numerous that they invaded chickenyards: "In spite of these almost incredible numbers, how short a time has it taken to despoil them even of this, the 'Garden of the West.' " He and other authors likened their passing to that of "the red-man": "Both step by step . . . have been driven from the haunts they once frequented, and over vast districts once all their own. They now retain possession of small and isolated spots, where they continue to look back with lingering longing over the fair domain, from which they have been ruthlessly driven." The inclusion of letters to the editor in periodicals invited individual observation from a local perspective of certain species' departure from settled land. A correspondent from Georgetown, Kentucky, in 1836 to the *American Turf Register* stated that the population of the common gray squirrel had greatly declined "from about the year 1825"; "from the first settlement . . . [it was] very abundant."[50]

Reading the Works: Individual Responses

Numbers of copies sold does not convey how the books were read, as historians of reader response are wont to demonstrate. Authors endeavored to convey meanings, yet each audience member may or may not have made the intended interpretations. Examples of individuals' and small groups' interactions with the natural history texts and derived textbooks indicate more solidly when and how this historical readership responded. Natural history discourse expressed an activity, language, and way of seeing that could absorb readers. These "respondents," both naturalists and non-naturalists, left compelling evidence of how they studied and even wrote in the natural history discourse.

The books themselves in precious instances testify to levels of study. Names on flyleaves suggest gender and age, such as the Parley *Botany* bestowed on a girl cousin. Similar evidence of ownership implies that girls as well as boys were reading the botanical and zoological textbooks. Some copies of natural history texts show an intensity of study and activity over the years. Botanical and ornithological texts, for example, were used as checklists to note sightings and holdings in collections, the owners often recording the dates and places they encountered a species over the years. Owners wrote in newer scientific names from other authorities to update these quasi-manuals. One copy of the *Ornithological Biography* indicates its role as a manual to identify specimens such as a snow bunting, "Shot near Lake Umbagog [Maine] May 26, 1862. I noticed more stripes & odd spots of brownish black on the belly." The Parkman family of Boston, close friends of Audubon, demonstrate that they read the *Ornithological Biography* for quite different purposes. George Parkman wrote the opening lines of William Cullen Bryant's *A Forest Hymn*, "The woods were God's first temples," in the margin next to a passage similar in sentiment in which Audubon describes a "lost" man praying for his soul on his knees in the wilderness. A family member bound to the Audubon page Mrs. Eliza Parkman's neat copy of Bryant's complete poem. The family was connecting two American Romantics' melding of the traditional functions of religion into nature worship.[51]

Because books carried the structure and substance of natural history, most adherents derived most knowledge from them. A range of material from children's textbooks to scientific monographs often triggered the fancy of individuals who would become naturalists. The new practitioners would build on their knowledge and fuel their desire by using the books to identify specimens and learn the specialized terminology. They, in turn, would foster the discipline by corresponding with other devotees and would sometimes write their own works. For the comparatively few Americans who persevered, books were the conduit to their favorite subject and activity.

Botanical textbooks far outsold zoological textbooks, strongly indicating that botany was the most widely practiced discipline of the natural sciences. The growing inclusion of botany in common schools, academies, and college curricula from the 1840s onward and the numbers of small yet active botanical clubs (often made up of women) support the conclusion that by 1875 tens of thousands of men and women probably knew the basics of anatomical structure and how to identify plants from these manuals.

Individual cases, however, give more definition of how botanical manuals affected lives. Asa Gray, whose own manuals would become classics, bought Amos Eaton's *Manual of Botany* in 1828 and remembered "poring over its pages, and waiting for spring." That spring, he "sallied forth one April day into the bare woods, found an early specimen of a plant in flower, peeping through dead leaves, brought it home, and with Eaton's 'Manual' without much difficulty [he] ran it down to its name Claytonia Virginica. . . . [He] was well pleased, and went on, collecting and examining all the flowers [he] could lay hands on." Thure Kumlien, the Swedish immigrant ornithologist, and the young Edward Lee Greene studied wildflower identities and Latinate nomenclature around Lake Koshkonong after Thomas Brewer had obtained a copy of "Grey's [*sic*] botany" for Kumlien. Greene packed Alphonso Wood's textbook when he went to fight the Civil War in Kentucky, Tennessee, and Mississippi. He wrote Kumlien of his new finds during the long trudges and even sent back specimens. In the midst of his soldiering and botanizing, he wistfully mentioned to Kumlien that he hoped to be back "next spring so we could gather Arethusa [an orchid] from that blessed little tamarack marsh" near Kumlien's home. Once he espied a new plant during a march but did not dare to break ranks; instead, he memorized the spot for later retrieval. Greene became a prolific academic botanist "with a wider field knowledge of the North American flora than any other botanist of his day." Anonymous readers not destined to become bot-

anists left vivid traces of their interest in the tentative pencil outlines
alongside the plant diagrams in Gray's works for the youngest readers. The next chapter discusses how women studiously adapted botanical illustration for their personal artwork.[52]

Other disciplines were not favored with such influential manuals
in these early decades. Part of the problem may have stemmed from
perceived difficulty in studying ornithology and mammalogy. Entomology held so many genera that decades of research were required
to prepare the simplified data presented in Asa Packard's *Guide to the
Study of Insects*, first published in 1869. Many shell collections contained mostly colorful foreign shells as opposed to the plainer inland
native species, so the few imported conchology works may have satisfied the market. For those interested in zoological subjects, natural
history monographs had to serve as introduction, checklist, and definitive authority, as the Audubon *Ornithological Biography* annotated
copy suggests.

To the relatively few eager pursuers, these texts proved a potent
introduction to fauna. For example, the detailed text explaining insect physiology, anatomy, and identification plus the exquisite wood
engraving in Thaddeus W. Harris's revised edition of *Insects Injurious
to Vegetation* immediately attracted the seventeen-year-old John Henry
Comstock in 1876 when he was directed toward the scanty science
book section of a Rochester bookstore. Another entomologist began
natural history with an interest in ornithology. Naturalist Samuel
Francis Aaron (1862–1947) sought for many years to find the scarce
1854 edition of Brewer's *Wilson's Ornithology*, for sentimental reasons:
his wife inscribed on his copy's flyleaf, "This was the work with which
S. F. A. first studied birds in Tennessee." Earlier, the future entomologist William Henry Edwards had so thoroughly absorbed Nuttall's
Manual and Audubon's octavo *Birds* edition as a Williams College
student that in 1845 he was moved to correspond with Audubon with
new information. The curiosity of the eminent naturalist William
Healey Dall, who studied under Louis Agassiz at Harvard, had been
piqued as a youth in Boston in the 1850s. According to one biographer,

> The accident that led him to become interested in shells was,
> he said, the possession when a boy of twelve of a copy of
> Dr. Gould's "Invertebrata of Massachusetts." Inspired by this
> work, and living near Boston, he undertook to make a com-
> plete collection of the shells of Massachusetts. Finding spe-

160

THE BOOK
OF NATURE
Natural History
in the
United States

cies that he was unable to name, he made bold to consult the author, Dr. Gould, who gave him much sound advice, and whom Dall characterized as "one of the best and most lovable of men."[53]

The cases of ornithologists Robert Ridgway and Frank Chapman demonstrate the young naturalist's progression from children's literature to scientific treatises. Ridgway, born in 1850 and growing up in a small Illinois town, Mount Carmel, remembered that the only books "that dealt, even in part, with natural history subjects that I had access to were an edition of Goldsmith's 'Animated Nature' (published by J. Grigg, Philadelphia, in 1830), belonging to Uncle William . . . and a History of the United States of equally early date [probably a Parley production], in which about 34 pages dealt with birds. . . . It was from this book that I first learned of Audubon, Wilson, Bonaparte, and Nuttall." Ridgway related the day his mother bought for him the handsomely illustrated and embossed edition of the Parley *Illustrated Animal Kingdom* at the sacrifice of all her spending money: "[T]his book with some of the front pages gone and many of the woodcuts disfigured by my efforts to color them is still one of my prized possessions." His first knowledge of technical terms came from a William Ruschenberger text, "of which I had somehow learned and bought from the publisher, the price being I think one dollar." Because these texts included no instructions for fieldwork, he had to learn to shoot and draw on his own and did not learn how to skin birds until he began working for the Smithsonian. In this isolation, until his early teens and contact with Spencer Fullerton Baird in 1864, "He had heard of the work done by the fathers of American ornithology, Wilson, Audubon, Nuttall, and Bonaparte, but did not know even the name of a living naturalist, and naively supposed that he was the only worker in birds in America." At the beginning of their fruitful relationship, Baird sent Ridgway his 1859 *Catalogue of North American Birds*, from which Ridgway learned the latest scientific nomenclature.[54]

Frank Chapman, born a few years after Ridgway, had a slightly broader range of early influences. He too vividly remembered being given the 1872 edition of *Illustrated Animal Kingdom*: "[F]or the succeeding ten years this was my only bird book." From his later perspective, however, he deplored its "complete disregard of the laws of geographic distribution." He remembered reading John Burroughs's 1872 *Wake Robin*, which he found expressed his "own response to the song of the Wood Thrush." By 1888, he owned the 1876 Porter &

Coates edition of Wilson's *Ornithology* and Edward Samuel's *Birds of New England*, and, when glancing into a bookstore window in New York City, discovered Elliott Coues's manual, the *Key to North American Birds*. Like Ridgway, "now I for the first time learned that there were living students of birds, worthy successors of Wilson and Audubon." In 1884 he began his long career of ornithological activities in earnest when he volunteered to note bird migration for the American Ornithologists Union after meeting a "professional ornithologist," Dr. A. K. Fisher, for he now had "a teacher, a textbook, and an object."[55]

The ornithological texts also could galvanize adults to become nature enthusiasts. Taxidermist Martha Maxwell, whose exhibits of Rocky Mountain animals impressed the Philadelphia 1876 Centennial attendees, wanted to learn the scientific background of the creatures she prepared. She studied her copies of *Birds of North America* (the 1869 catalog by Baird) and James G. Cooper and George Suckley's *Natural History of Washington Territory*, which covered all fauna in the region; however, she complained to Baird, to whom she had sent some specimens, that such authors expected of their readers classical education in Latin and so did not include "the definitions of scientific terms. Webster's Unabridged is my only assistant in my perplexing research in this direction." (Baird pointed out Coues's new glossary of technical terms, available in several publications.) James Belden, another Baird correspondent and early supporter of the Cooper Ornithological Club in California, who had hunted ever since he was a boy in rural Massachusetts, became interested in identifying birds in his forties when he lived in California. The catalyst was a book: "Early in the spring of 1876 I got a volume of *California Ornithology* and began industriously to collect and identify the birds of the State." He expressed wonder at the new focus and knowledge opened to him: "I had been an ardent sportsman ever since I was a small boy and I supposed that I knew most of the birds, but my first bird book astonished me with many I did not know and had never heard of. I had never met an ornithologist or oologist and did not know there was any in this State." Belden admitted modestly that he became successful in identification thanks to Baird's volume of the Pacific Railroad reports and "Wilson's simple descriptions." To both Belden and Maxwell, Baird offered more books for only the price of postage, that they might continue their efforts.[56]

Texts thus created the networks of correspondents, which in turn fed further texts. They could also stimulate gatherings of the "most ardent lovers of Nature" (to use a characterization by Martha Max-

162

THE BOOK
OF NATURE
Natural History
in the
United States

well) in an area. Kumlien and his former student Greene, consulting
their botanical and ornithological manuals, formed one such cluster
in Wisconsin. William Strain, the postmaster of Greensboro, South
Carolina, could avidly read every detail of the golden eagle in his
congressman's copy of the octavo *Birds*. The brilliant scientist, artist,
and porcelain collector Edward Sylvester Morse and his close boyhood chum, John Gould, in Portland, Maine, chose to study the land
snails, according to a candid biographer, "the smallest, ugliest types"
of shells. He and Gould by the age of twelve began their local collection; they acquired Augustus A. Gould's *Report on the Invertebrata of
Massachusetts* and borrowed other titles from acquaintances. One
diary entry recorded when they were both young men reported:
"Went down to John Gould's in the evening; carried some of my Shell
books with me. Read to John on the Lymnidea." William Brewster
and Henry Wetherbee Henshaw had known each other as schoolboys
growing up in Cambridge, Massachusetts, in the 1860s, with Brewster's lessons on stuffing birds cementing the friendship. As mentioned in the introduction, these two met in 1871 on Monday
evenings in Brewster's snug attic study to read his father's copy of the
Ornithological Biography, and soon other "like-minded" friends, such
as future Audubonia collector Ruthven Deane, joined the discussions.
In 1873, the group formally organized itself into the Nuttall Ornithological Club, named after the ornithologist who had mentioned the
Cambridge area frequently in his *Manual*. The group continued to
discuss other writings (one member, Henry Purdie, was particularly
noted for discovering obscure sources in the local Harvard and Boston libraries) and published the members' original articles in their
own magazine beginning in 1876.[57]

These individuals' experiences attest to the expense and rarity of
the books. For example, Kumlien, Maxwell, and Belden could not
obtain locally any texts and had to rely on acquaintances like Brewer
and Baird to supply them. Isolation and scarcity of information makes
Kumlien's plaint to Brewer additionally poignant: "I have not seen
anybody that takes any interest in anything else but wheat, potatoes,
and corn. You do not wonder if I am greedy for books." To receive
his edition of Brewer's Wilsonian ornithology in 1848, Kumlien made
complicated arrangements with a friendly minister to buy it in Milwaukee for three dollars. Ridgway noted that he did not find the Ruschenberger volume on a bookstore shelf but was required to order
it. The John Henry Comstock story mentions the lack of other scientific texts in the 1876 Rochester bookstore. Even John Bachman, writing the *Quadrupeds*—the definitive work in its field in a major U.S.

natural history center—continually asked the Audubons in New York
to send European references to him. They, in turn, needed to ask
colleagues in Boston and Philadelphia to copy articles Bachman had
encountered in synonym listings but had not yet read (one Boston
friend asked the Audubons to be careful with the book he had wran-
gled out of the Harvard Library).[58]

Cost further prohibited use; Comstock had to ask for a substantial
advance on his salary to pay ten dollar for *Insects* and Brewer mentioned that even a secondhand copy of Gray's *Manual* could cost two
dollars in 1848. Kumlien could not hope to buy the only currently
available work on mammals, the thirty-dollar octavo edition of Audubon and Bachman's *Quadrupeds*, although he wanted one to pursue
his interest. Ridgway remembered that the elaborate Parley *Illustrated
Animal Kingdom* cost between nine and ten dollars. Book historians
have questioned how many could afford books when the cheapest
cost a quarter and the average wage was a dollar a day. The possibility
of buying science books would have been far more remote. Libraries
were only beginning to build collections of their own to improve access. (Henry David Thoreau was able to identify a bird using the octavo *Birds* in the Concord Public Library in 1855.) The young
Spencer Fullerton Baird may have undergone the greatest physical
effort to borrow library books on ichthyology, paleontology, and invertebrates when he walked round trip from Carlisle to the Pennsylvania State Library in Harrisburg, along the railroad tracks—a journey
of roughly eight hours—all in one day.[59]

The scarcity led to an intensity of the reading experience, as
these enthusiasts pored over their hard-won copies. Dall copied passages from the conchological works he borrowed for his future reference "because he thought he could never buy them." Chapman
eventually could identify birds with his Parley book "in his head."
Comstock virtually memorized the richly detailed *Insects*, and Henshaw remembered feeling like an ornithologist after memorizing the
Latinate names in Baird's catalogs of North American birds. Morse
and his friend Gould learned to distinguish minute differences in
their shell collections by deciphering taxonomic descriptions that
often were unillustrated. Maxwell rose to the task of reading the un-
yielding Cooper and Suckley works, full of extensive nomenclature
and life histories challenging to a beginner. Brewster and Coues had
so often read Audubon that, consciously or unconsciously, they imitated his dramatic style intended to engage the reader's attention;
Coues's rich description of the canon wren quoted in the last chapter
is a case in point. Using the preeminent texts as introduction posed

164

THE BOOK
OF NATURE
Natural History
in the
United States

formidable obstacles but ironically held the rewards of thorough knowledge of the nomenclature, physiology, and life histories of many species for those intelligent and dedicated enough to continue.[60]

Incorporating the Texts

Not all readers who at one time enjoyed textbooks and participated in natural history activities or built collections went on to become noted scientists or active "amateur" collectors like Belden and Maxwell. Other interests superseded their youthful hobbies, yet their biographies reveal the special role of natural history and its books at certain times in their lives. Knowledge of natural history could bring sustained enjoyment or provide a focus during a given period.

Daniel Chester French, for instance, with his close childhood friend William Brewster, learned how to stuff birds from his father, a distinguished Massachusetts jurist, who owned a copy of the Nuttall manual. Young French, like his father, kept a "day book" of birds, focusing on the first sightings of the season. The ten-year-old noted in the tables he carefully drew and lovingly decorated whether the bird had come singly or in pairs and what kind of nests and eggs it had. His attention to one aspect of the natural world that showed its yearly cycles is reminiscent of the earlier *flora calendrae*. This exacting approach, smacking of both the businesslike and the romantic, reattached the bird observers to the slowly receding unspoiled natural world. French lost interest in ornithology and gave his sizable egg collection to Brewster, deciding instead to develop the artistic bent that carried him to the apex of Beaux Arts sculpture.[61]

Natural history was prescribed in literature as a morally uplifting yet interesting way for youths to pass time in a society fraught with temptations. As Thaddeus Mason Harris told Harvard's Natural History Society, "He who loves nature, loves not revelry; artificial excitement has no fascination for him. The overflowing cup and unmeaning and dishonest game, cannot entice him." Some examples show that youngsters did, in fact, passionately enjoy the hobby and spent hours honing their craft. The young William Rollinson Whittingham grew up to be fourth bishop of the Episcopalian diocese of Maryland and a leader of the church. His mother, a woman of formidable character and a tremendous influence on her firstborn's life, taught her son how to write at two years, and "French and Latin and Greek . . . were all entered on at an age when other children are in the nursery playing with alphabet blocks." No frivolous hobbies were allowed for this boy, but the study of nature—including mineralogy, physics, and especially botany—interjected outdoor activity, interesting objects, and visual beauty into an otherwise demanding routine. He drew his mineral specimens and as a teenager delighted in taking long walks. One anecdote tells of the lasting pleasure (and pride) his botanical knowledge gave him. On one visitation, the bishop corrected the owner about how many varieties of trees edged his driveway and, actually seizing his host's arm "with that impulsive energy so characteristic of him," propelled him outside to point out the very specimens. When asked how he noticed so much, he reiterated the familiar rationale for natural history study, which he had truly incorporated: "My mother always told me that God had given me eyes to see with and observe."[62]

Thomas Wentworth Higginson, abolitionist, leader of black troops in the Civil War, confidant of Emily Dickinson, and man of letters, grew to intellectual maturity in an atmosphere different from that of Whittingham's Calvinist-flavored youth, that of Unitarian Boston, yet he too spent his youth "with butterfly-net and tin botanical box." Even in his busy days as a young minister, he pursued microscopy and helped organize the Worcester, Massachusetts, Society of Natural History, having previously enjoyed "field study" walks with Harvard entomologist Thaddeus W. Harris. At seventy-two years old, he reminisced about the pleasures that "early acquaintance" with flowers, birds, and insects via natural history studies had brought him throughout his life: "Every spring they come out to meet me, each a familiar friend, unchanged in a world where all else changes; and several times in a year I dream by night of some realm gorgeous with gayly tinted beetles and lustrous butterflies." Natural objects, by their disappearance, marked unwelcome change as well: "I associate with each ravaged tract in my native town the place where vanished flowers once grew,—the cardinal flowers and gentians in the meadows, the gay rhexia by the woodside, and the tall hibiscus by the river." In an especially delightful passage, he joked that "nominally botanizing" was an ideal vehicle for a moody youth's self-reflection. At twenty he had forsaken plans for a law career but had not decided his future path. He spent much of that summer pursuing the *Andromeda poli-folia*, a celebrated waterlily, in Hammond's Pond in Brookline, Massachusetts, and fancied himself in one of his favorite German romances, for "my lilies were as pale and as abundant as any German lake could ever boast. . . . I spent much time in the woods, nominally botanizing but in reality trying to adjust myself, being still only nineteen or twenty, to the problems of life."[63]

Susan Fenimore Cooper studied natural history texts in order to

166

THE BOOK
OF NATURE
Natural History
in the
United States

understand better the inhabitants of a special place—the grounds and surrounding woods of the Cooperstown, New York, estate of her father, James Fenimore Cooper. She drove or walked the grounds twice a day in all but the coldest months, noting the seasonal changes through the foliage, progression of bird nesting, and appearance of wildflowers. The enforced time indoors was spent reading ornithological texts and researching local animals such as moose and trout through the assistance of James DeKay's *New York Zoology* (she also asked Victor Gifford Audubon for moose aid) in order to learn habits, localities, and related information beyond her own observations. This material gave Cooper an enhanced familiarity with such local species as the loon and water dipper because she had read about related species in the American West and in Europe, thanks to Audubon and Charles Bonaparte's continuation of Wilson: "Mr. Charles Buonaparte [*sic*] mentions having frequently watched them [the dippers] among the brooks of the Alps and Apennines." Her *Rural Hours* (1851), composed as a diary commenting on the scenery in her world, melds empirical information derived from outside sources with her personal meditations on the landscape. Her enlarged knowledge of the natural world imparted by natural history texts led her seemingly effortlessly to include for her readers connections between the local flora and fauna and those of the entire world, as when she comments that a particular flower is "in common with Northern Asia," which was one of Asa Gray's observations.[64]

Writing Natural History

It already has been suggested that reading natural history texts influences, if not dictates, the writing of natural history. The young Brewster and Coues, for example, followed the style of the popular Audubon, and William Strain gave physical dimensions of his golden eagle like Audubon's description in the *Ornithological Biography*. After reading his crow history, admirers related to Audubon anecdotes of their own thieving crows snatching away marbles. Susan Fenimore Cooper listed common birds and their appearance for her readers, just as her natural history references did. The practice and language of natural history reinforced a way of ordering the world that focused on species, with the individual encountered representing the whole. The characteristics of appearance, locality, and habits that distinguish it from like species receive primary attention. Passages often list physical identifiers such as size in numerical figures, include Latin, and, in the longer prose sections, adopt the third-person narrative.

All persons writing in this style about natural objects may be presumed to have been exposed to the discourse, although their results may be rough compared to naturalists' prose. For example, M. V. B. Morrison, in his autobiography, *The Orphan's Experience: or the Hunter and Trapper*, published in Des Moines in 1868, wrote this description of a pouched rat (which Morrison called a "gopher"), in which he portrays the many individuals of a species as one in appearance and habits, resembling natural history phraseology but without Latinate terminology: "The gopher is an animal that burrows in the earth, and resembles, to a great degree, the mole, only it is much larger— probably as large as a full grown rat. They are of peculiar formation, having what might be called a pocket on each side of the mouth. These they fill with dirt as they dig their way through the earth, and when filled, the gopher will come to the surface, turn a complete summerset [*sic*] for the purpose of emptying the pockets, then disappear." One may speculate that the author had read similar passages applied to animals from childhood onward and so automatically adopted the style. Noting such usage in "non" natural history writing, such as this autobiography, further reveals the impact of the discourse.[65]

Travelers, too, used the language of natural history to make sense for themselves and to describe to others the often confusing new worlds they confronted. Nineteenth-century Americans were particularly exposed to changing territories as they moved between Old and New Worlds and from the East to the new West. For example, the distinguished Boston physician John Collins Warren, author of a scientific monograph on the mastadon and subscriber to Audubon's works, displayed that particular focus and diction in his comments on flowers and birds on Hampstead Heath in his unpublished England travel journal. "Birds. The thrush most various in its notes. Skylark's notes very fine. sings on wing.—Nightingales sing in the spring only— Plants. Steatus [?] on commons. Goosebriar. Belladonna etc. in hedges &. Grows in the chalk soils. The chalk formations contain the greater number of animal remains."[66]

Natural history discourse held significant psychic force in the adaptation of immigrants to their new home. Henry Moore, a Baltimore merchant who moved to Kentucky in 1807, avidly explored caves but paid special attention to the plants around him with which he was unfamiliar. In his diary he spoke of the individual plants as one entity upon each sighting and listed physical distinctions ("The plant climbs, three heart-shaped leaves, etc"). His search for and careful listing of the individual plants evidently gave him great emotional satisfaction and accustomed him to his new surroundings. The rec-

168

THE BOOK
OF NATURE
Natural History
in the
United States

ords of each plant's yearly blossoming and fruition additionally enhanced knowledge of his home, as he could then measure the progress of future seasons by observing the flowers. Thus, the framework of natural history discourse and practice could provide stability in an initially uncertain place and time.[67]

The prerequisite for this ordering of new surroundings was some prior knowledge of natural history. We do not know where Moore picked up his natural history diction, whether from medicinal herbals still in use in his boyhood years or botanical textbooks. In fact, relatively few Americans employed Latinate terms or general natural history diction in their western travel journeys but much preferred to praise the scenery in general romanticized rhetoric. One diarist enthused: "A beautiful spring morning, our first experience of the almost magical change which in a few days comes over the face of Nature. The freshness of every-thing had lured my sister and myself two miles. . . . The flowers were coming forth all around us." She is content not to list the plants.[68]

One group of Americans, however, the government surveyors, was charged with scientifically describing the new lands on behalf of their fellow citizens. Most expeditions had specially appointed naturalist-artists and naturalist-surgeons, whose first experience with intensive collection and description was often the surveys. Despite their limited previous knowledge, they dutifully measured and entered in notebooks the specimens collected and mentioned the flora and fauna in their personal daily journals. Majors William Emory and John Frémont, leaders of their expeditions, did not scorn to collect botanical specimens on their Southwest expeditions and to record their plant specimens in diaries. The young Caleb Rowan Kennerly, attached to the northern route of the Pacific Railroad expeditions under Governor Isaac Stevens, apparently had little prior experience in natural history (for instance, he was not highly skilled in its Latinate nomenclature), but he persevered in capturing specimens such as the scorpion in his bedclothes ("we soon produced a vial, & capturing our enemy imprisoned him securely in it, & removed forthwith to another apartment, not to sleep sweetly but to dream of poisonous insects, snakes &c all night"). He falls into the natural history style when describing a specimen: "[T]his morning I succeeded in killing some Mexican partridges (*Callipepla gambella* [a notation evidently added later]). They are somewhat larger than the bird of that name in the states & very far surpass them in beauty. They are generally of a dove color, but beautifully marked on the sides, wings & head, & the latter in the male is crowned by beautiful plumes. Their flavor we can better describe after dinner."[69]

George Suckley and James G. Cooper, two youthful but proficient naturalists who also served on the railroad expeditions and the 1857–1861 Northwest boundary survey, carried along Audubon's *Synopsis* and Baird's catalogs of reptiles to confirm the identifications and indicate new finds ("Saw here the lizard No. 11 of Coll. 1853 among the logs"). Suckley's diary neatly picks out the salient characteristics of the animal: "A large serpent called the 'prairie bull snake' was caught and brought to me, it appears to be harmless, & although it will, when irritated, try to bite, its mouth is destitute of poison fangs—Length 60 inches—belongs as nearly as I can place it to family Coluber Genus Pituoplus[?]" In the process of acclimating to his new surroundings, Cooper noticed that the date of blooming of the familiar *Trillium grandiflora* differed from that in his home back east. Anecdotes were less frequent but were recorded: Dr. George Hammond of the U.S. Army at Fort Kearny communicated the interesting story of his discovery of rawhide rope in a wolf's intestines. Lighter moments, such as one railroad expedition group's making an American flag from muslin and red and blue flannel undergarments belonging to Suckley and their leader, Governor Stevens, to "appropriately celebrate" Independence Day (one member becoming "gloriously drunk"), did not make the final report.[70]

An anonymous, unemployed New Englander joined a canal boat crew to journey on the Chesapeake and Ohio Canal in 1858. Roughly thirty years later, he wrote about the way of life in a lovingly penned and illustrated manuscript memoir. He freely speaks of his frequent "botanizing and geologizing" and was especially interested in both river and land shells. His text gives the physical listing of shells' appearances (for example, "a smooth cone about one-half inch in diameter and the same in length"), but he devotes his most complete descriptions to insects, particularly to their nest building. One afternoon he observed tumble bugs rolling some horse manure on the road to holes in the field for future larvae to eat. His description relies on the representative type's actions to stand for the entire species: "The female went behind the ball and placing her front and middle feet on the ground with her head down pushed and guided the ball with her hind feet." The process aroused respect for the beetles' strength: "I saw them in all the different processes, the forming of the ball from the fresh manure, the roll across the road, the exciting tumble down the ditch, which might be compared to two men rolling down a fifty-foot cliff clinging to a barrel, then all the various stages of climbing up the bank, a mountain in size to them, and rolling back down again, the journey through the field and the digging of the pits where they were buried." After the captain had pointed

170

THE BOOK
OF NATURE
Natural History
in the
United States

out a swarm of mud wasps making a colony on a cliff, the author observed the females making balls "of one eighth of an inch in diameter" into tubes "of two hundred and fifty to three hundred clay balls" in which she laid the eggs. The sentiment the writer expresses ("when he has studied their work [he] is inclined to question the idea that only man has reason") harkens to the 1840s and 1850s anecdotes extolling animal intelligence.[71]

The writer deplores the "idea of natural history" expressed by Pic, a free Black who was his canal associate. Pic believed that frogs rained from the sky despite the writer having explained the tadpole's metamorphosis, leading the writer to conclude that, "Whenever a person has made up his mind that something is produced by some supernatural cause, no amount of argument or proof that it was produced according to the common law of nature will have the least effect with him," a lament reminiscent of Wilson's earlier derision of superstitions. The author also remembers that "the country people" outside Cumberland thought "I was crazy . . . with my erratic zigzagging from creek to rock and from rock to wild flower."[72]

This anecdote demonstrates that natural history perspective or practice had not spread into every American life. Indeed, it remained distinct from popular and vernacular knowledge and activity. The circumstances of antebellum and postwar American culture made the discourse known to a larger number of individuals than ever before, however, and available for use by such persons as the C&O canal drifter as well as the well-known editor Thomas Wentworth Higginson. Periodical literature and textbooks introduced the subject matter to both the young and adults; the few who were so stimulated by the subject as to learn more derived their knowledge from textbooks and major monographs such as Audubon's. Those who found emotional and spiritual satisfaction in other pursuits could still employ the writing style for describing new environs and, as in Susan Fenimore Cooper's case, familiar haunts. Through the act of portrayal, they became more assured in the natural world. Youthful acquaintance with nature acquired through study could provide pleasure throughout a lifetime, as in the cases of Whittingham and Higginson.

The greater availability of natural history literature accelerated the making of scientific monographs. In earlier times, Catesby had read Lawson, and Bartram in turn made use of Catesby, with only a portion of the new land discussed. Now the increase in sources dispersed over a great geographical area stimulated new devotees who could investigate the still little known regions. Westerners Martha

Maxwell and James Belden contributed specimens to Baird's Smithsonian efforts after they had been inspired by other monographs. Thure Kumlien and Edward Greene were able to botanize around their Wisconsin home; Greene, aided in part by botanical manuals, proceeded to a distinguished career publishing California species. Audubon and Wilson, as (freely) interpreted by "Peter Parley" and other textbook writers, inspired Robert Ridgway and Frank Chapman, the next generation of ornithologists. The great economic and technological forces underlying the print explosion influenced both individual lives and discipline development.

Dissemination encouraged a new familiarity with flora and fauna, native and foreign, different from the daily experience. Virtual "biographies" in flower books, periodicals, hunting literature, and textbooks elaborating existence from birth to death gave the subjects a weight and presence not unlike that accorded persons. The authors' intent to cherish native flowers and to protect wildlife introduced generations to the concepts involved in preservation. The descriptions, in addition, contributed masses of empirical information about habits and appearance. Yet the knowledge would have been incomplete without the potent visual images linked to the texts.

5

Dissemination and Response:
Visual Imagery

*I*n 1850 the family and friends of
Benoni Pierce presented the twenty-eight-year-old farmer of Pawling,
New York, with a quilt to celebrate his upcoming marriage. Eighty-
one squares, signed by the individuals who made and donated the
squares, compose the quilt. Each bears an appliquéd design of a subject considered beautiful or important, like a Bible, a bouquet of
flowers, or butterflies. In one square, an animal, readily identifiable
as a deer, turns its head and gazes past the viewer (Illustration 5.1).
Two abstracted trees represent its surroundings. The designer, Julia
G. Burdick, adapted the deer from a printed image ultimately derived
from Thomas Bewick's *History of Quadrupeds,* originally published in
Newcastle-on-Tyne in 1790 (Illustration 5.2). This chapter explains
how and why a female upstate New York quilter, half a world and half
a century removed, came to use this Englishman's wood engraving.
Books and periodicals, pouring out in unprecedented numbers in
antebellum America, were disseminating designs until then limited
to original works. Individuals adapted this base of mass-produced im-

agery in a variety of media to express their relationship to the natural world.[1]

Text and visual imagery worked closely in these "quotations." Authors and editors adapted both text and images for the same productions, so the dissemination of the visual often followed the same paths as those of the texts suggested in the last chapter. For example, textbooks and periodical literature joined wood engravings and text excerpts. The image and text retain the physical and intellectual association of the natural history books from which the articles were adapted. Individuals aping natural history description in journals, whether for private use or in government surveys, linked the visual to their verbal descriptions.

Visual imagery, however, involves methodologies and sources different from those associated with textual material. The translation

Illus. 5.1. Detail of quilt made for Benoni Pierce, by Julia Burdick, 1850, Pawling, N.Y. Number T.16323. Gift of Adelaide Pearce Green and Mira Pearce Noyes Boorman. (Courtesy of the National Museum of American History, Smithsonian Institution.)

174

THE BOOK
OF NATURE
Natural History
in the
United States

of original illustrations to wood engravings and lithographs implies changes in focus due to qualities inherent in the media as well as conscious adaptations by artists and engravers/lithographers and editors' contributions.[2] "Viewer response" to natural history illustration must be interpreted not only from letters and reviews but also from the treatment of the physical objects. Primarily visual media, such as handicrafts, drawing, painting, sketching, and decorative taxidermy, engaged in themes common to natural history and in certain cases (such as the quilt square shown in Illustration 5.1) "quoted from" actual prints to recreate parts of nature. These artifacts, not always regarded as historical evidence, suggest the aspects viewers deeply appreciated in natural history.

Patterns of Dissemination

Americans outside the charmed circles of subscribers and their associates did experience the images of Audubon and other naturalists, albeit in the form of designs adapted from the monographs. Because of new printing technologies and publishing practices, these designs, generally wood engravings, appeared in popular natural history and the other genres involving life histories, including children's and hunting literature. Wood engravings produced in the 1830s pro-

Illus. 5.2. "Stag." Wood engraving by Thomas Bewick, for Thomas Bewick and Thomas Beilby, *A General History of Quadrupeds* (Newcastle-on-Tyne, 1790), 105. (Courtesy of the Department of Special Collections, Van Pelt–Dietrich Library Center, University of Pennsylvania.)

vided a stable of images for the vastly popular Peter Parley works and
the Harper Brothers' Family Library series, which, with addition from
English imports, were reused in periodical literature and other publishers' output into the 1860s. The bulk of images favor the "living
animal or plant" mode of representation over the diagrammatic or
taxonomic. These recurring images introduced to a larger audience
than ever before a variety of native and foreign creatures portrayed
in a more accurate manner. Because this audience included artists,
designers, and craftsmen, the natural history figures influenced conventional representation of certain types of animals.[3]

The Use of Wood Engravings

Few natural history illustrations outside major monographs were
available even to the wealthiest Americans before 1825. The various
imports of Goldsmith's *History of Animated Creation* and the illustrated
encyclopedias and the American editions of the same held the greatest number and highest quality of such engravings. The difficulty and
expense of executing original sketches and engraving, etching, printing, and hand-coloring plates prohibited magazines from offering
many examples to readers. The few periodical prints of animals published in the early republic show an uneven quality in conception
and execution, ranging from an adequate representation of a moose
derived from a striking George Stubbs design to a cruder, humanoid
image of a buffalo (Illustration 1.6). Children's literature often relied
on the cheaper media of metal and woodcuts, which permitted only
simple shapes, as in Noah Webster's 1812 *Zoology,* which contained
only small woodcuts of basic animal forms pasted in the margins.[4]

The technique of wood engraving, revived in England by Thomas
Bewick in the late eighteenth century, would prove a cheaper process
than metal engraving but still capable of rendering the necessary detail. Wood engraving, basically the cutting of linear designs with engraving tools across the grain of the wood, as practiced by Bewick
captured not only precise markings of birds but also the characteristic
outlines of large animals, as in the stag (Illustration 5.2). It could also
suggest backgrounds—all in an astoundingly small area, often two
inches by three inches. The blocks could be printed on the same page
as letterpress and withstand thousands of impressions, thus defraying
costs. Bewick prided himself on drawing native species from life, but
ironically the exotic animals he copied (mostly from Buffon's *Histoire
Naturelle*) were extremely popular. The "father of wood engraving in
America," Alexander Anderson, had taught himself the craft by copying Bewick's great work, *History of Quadrupeds,* in 1797 and reen-

176

THE BOOK
OF NATURE
Natural History
in the
United States

graved both that work and Bewick's *British Birds*. Anderson himself taught the next American practitioners of the technique, including John H. Hall, who would later engrave for Nuttall's *Ornithology*.[5]

Changes in the publishing industry would more fully exploit the potential of wood engraving in the decades following 1825. The emergence of major publishers with nationwide distribution networks and the invention of the stereotyping and electrotyping processes that would make wood blocks even more durable made the works of Bewick's American disciples accessible on a scale unimaginable in the early republic. Stereotyping, first used in the 1830s, and the later electrotyping method permitted the creation of metal replicas of the wood-engraved blocks and so enabled press runs of tens of thousands of images.

A group of Boston wood engravers, possibly led by Abel Bowen, engraved the first and largest group of natural history images destined for intensive "recycling" during the next forty years. They illustrated an 1831 American edition of the Englishman Oliver Goldsmith's *History of Animated Nature*, edited by John Wright earlier that year. (The original London edition itself had relied on copying Bewick's dogs, deer, and opossum.) The anonymous American editor who copied the text and "over four hundred" illustrations selected "the best engravings in the London edition, but we have added many others from the Zoological Garden [and] Tower Menagarie [publications]," and promised to include "American zoology" in the form of "Wilson and Bonaparte." The London-produced *Tower Menagerie* and *Zoological Gardens,* which as promised were heavily drawn upon, featured the designs of British artist William Harvey as translated by the wood engravers Branston and Wright. According to the author of both works, Edward Turner Bennett, Harvey had studied the "manners" of live foreign species in both London menageries and so "invested their portraits with that natural expression in which zoological drawings are too often deficient." The vitality of Harvey's figures, drawn in animated poses, persisted in several American publications (Illustration 5.3). The Boston engravers, including Bowen, J. C. Crossman, and the budding landscape artist George Loring Brown, reengraved the London images with few changes and added many images from Wilson and Audubon's first volume of the *Birds*. Wilson inspired the majority of bird engravings, but they memorably adapted the Audubonian "bird of Washington" (later proved to be an immature golden eagle), the mockingbird attack, and the male wild turkey (Illustrations 5.4 and 5.5). Exotic species were not neglected but celebrated, as evidenced by the eight separate engravings of the elephant.

In striking contrast to the well-illustrated mammals and birds, the
invertebrates are figured with only one image per loosely defined
group (i.e., bee, sphinx butterfly).[6]

Samuel Goodrich's "Peter Parley" publishing concern used these
same images for the *Book of Ornithology* and the *Tales of Animals* with a
few redactions of the text. The wood engravings, either the same or
duplicate blocks, also illustrated the Parley's adaptations of Goldsmith, *The Naturalist's Library,* which was issued from 1833 to 1851
(the latter edited by A. A. Gould), and *Natural History Abridged for the
Use of Schools by Mrs. Pilkington,* from 1833 to 1857. Goodrich's extensive oeuvre featured the images in other genres, and his practice of
renting out plates facilitated the images' appearance in works he or
his editors did not guide. In numerous highly successful geography
and history works on the United States, he and others used the animal adaptations to promote the country's natural productions. Engravings showing trees' flowers and branches—adapted from one of
the few original works on American trees, François André Michaux's
North American Sylva—copiously illustrate the native forest descrip-

Illus. 5.3. "Jaguar." Wood engraving by anonymous artist after design by William
Harvey, for Augustus A. Gould, ed., *The Naturalist's Library* (Boston: Phillips,
Sampson, & Co., 1849), 185. (Courtesy of the Department of Special Collections,
Van Pelt–Dietrich Library Center, University of Pennsylvania.)

Illus. 5.4. "Great American Cock." Hand-colored engraving by Robert Havell Jr. after John James Audubon, for Audubon, *Birds of America,* vol. 1, Plate 1. (Gift of Mrs. Walter B. James. © 1997 Board of Trustees, National Gallery of Art, Washington, D.C.)

tions (these same plates appear in his botany textbooks). In 1858 another group of wood engravers, led by Augustus A. Fay, reengraved many of the Audubon *Birds* in a larger and more ambitious format for the Parley production team, resulting in *Illustrations of the Animal Kingdom* (another publisher bought the plates from Goodrich and retitled it *Johnson's Animal Kingdom*).[7]

The Boston firm of John Hall and Abel Bowen adapted the Wilson and Audubon plates to illustrate Thomas Nuttall's *Manual of Ornithology* (1832, 1840). They reengraved the Parley images and included new ones, mostly from Wilson. Their admiration of Bewick (Hall and Bowen, in fact, named their short-lived publishing firm the Boston Bewick Company) led them to include copies of his British raven and marsh titmouse for the American analogues. Bewick also informs their interest in precise rendering of features and overall movement. These same wood engravings were utilized in a variety of publications hitherto unrealized by historians, including John Lee Comstock's zoological textbook series and *Godey's* and *Graham's* mag-

Illus. 5.5. "Wild Turkey." Wood engraving by George Loring Brown after design by John James Audubon, for Samuel G. Goodrich, *Peter Parley's Book of the United States*, 102. (Courtesy of the Department of Rare Books and Special Collections, Princeton University Library.)

180

THE BOOK
OF NATURE
Natural History
in the
United States

azines. These plates were also reused in children's books of birds, such as the 1855 *Birds of the Woodland* illustration of a whippoorwill, which appeared as late as 1867 in another children's book published in Macon, Georgia (Illustration 5.6).[8]

The other textbook creators similarly adapted images from zoological texts into their book series and "recycled" them freely through their editions. The bird and insect works in the "Natural History" section of the Harper Brothers' Family Library series, for example, copied the English editions' adaptations of sources as varied as Audubon (his Baltimore oriole nest), Wilson, Françoise Vaillant's *Oiseaux d'Afrique,* and even the seventeenth-century illustrator Sybilla Merian (her hunter spider with bird in its grasp). Harper Brothers reused many of these bird and insect architecture diagrams in the popular *Uncle Philip's Conversations* series, children's books that relied on dialogue to impart messages. In the 1860s the college natural history teacher, Sanborn Tenney, summoned a crew of talented artists and wood engravers, among them Henry Marsh and Edward S. Morse, to adapt more recent American monograph imagery for his series of zoological textbooks. His introductions proudly emphasized

Illus. 5.6. "Whip-Poor-Will." Wood engraving, probably by Hall and Bowen, after design by John James Audubon, for *Birds of the Woodland,* 23. (Courtesy of the Department of Rare Books and Special Collections, Princeton University Library.)

the great care that he and his team gave to the electrotypes of "over five hundred engravings" his artists created from "rare books" at the Harvard, Boston Society of Natural History, and Massachusetts State Libraries, including Audubon's *Birds* and *Quadrupeds,* Wilson, Say, Holbrook's *Herpetology,* Binney's *Terrestrial Air-Breathing Mollusks,* Harris's *Injurious Insects,* Edward Dana's works on corals, and Edward Verrill's on worms. He gave "great prominence" to the "Mammals and birds of this country," so that all might have at least an "accessible catalogue of these two groups in which everyone is interested." The "other classes, however, are not neglected," and indeed, because of the specialized monographs of Harris and Dana, the invertebrates were well represented compared to the earlier Parley productions. Tenney and Abby Amy Tenney recycled the blocks through seven editions of his *Manual of Zoology* and her series designed for younger children.[9]

Botanical books followed the trend to adapt imagery from monographs, although Alphonso Wood claimed he drew all-new illustrations for his *Class Book of Botany.* Asa Gray again used his draftsman, Isaac Sprague, to make hundreds of diagrams for the manuals, many based on Sprague's work for Gray's other scientific publications. Lincoln Phelps's *Familiar Lectures* relied on Eaton's text and diagrams and acknowledged using copies of eight wood engravings from the French publication *Eléments de Botanique* to demonstrate climatic "characters" of plants. Her later editions adapted Gray's *Genera* plates, drawn by Sprague, to show the newer "natural orders" of plants, thereby exposing tens of thousands to those elegant designs.[10]

Further Recycling

Periodical literature boomed after 1830, but natural history images were limited to the few major periodicals specializing in illustrations. The magazines' editors did not commission original artists' work but borrowed blocks of images or had their engraver reengrave others' designs to an extent generally unappreciated. Thus, *Arthur's* and *Graham's Magazines* featured the "Whip-Poor-Will" (Illustration 5.6) and other birds from the Hall and Bowen wood engraving blocks of Nuttall's *Manual* to accompany the abbreviated life histories. *Gleason's* (later *Ballou's*) *Drawing Room Companion* made natural history illustrations a staple in the 1850s, with many wood engravings on such subjects as the iguana, the life history of the silkworm, and exotic shells (one vignette of underwater shells ran again within six months in *Godey's*). The editors ran whole pages of various "Curious Animals," "Illustrations from Entomology," and "Illustrations from Or-

182

THE BOOK
OF NATURE
Natural History
in the
United States

nithology," with the many lively species retaining their own characteristic backgrounds. Although the editors boasted that their artist (possibly William Croome, who also drew sporting scents) "drew from life as Audubon and Wilson did," the images are mostly derived from English wood engravings, most notably from the English publisher Charles Knight's popular natural histories.[11]

Hunting literature, like the other genres dependent on life histories, showcased these very same images, sometimes via the same wood engravings, or demonstrated the visual influence of Audubonian composition. Compilations of hunting stories intended to please the armchair traveler recycled game mammal images from the Parley productions. As late as the 1870s, Thomas Bewick's bull image (which wrongly but frequently doubled as an American bison) and the bears originally in Harvey's menageries of four decades earlier still roamed in tales of the hunt. In contrast to these well-worn borrowings, William Henry Herbert boasted that he himself drew most of his designs on blocks or copied others' scientific drawings to familiarize his audience with them. In his *Game Fish* he noted his artistic debts to Thomas Yarrell and Sir John Richardson, two English naturalists, and mentioned borrowing specimens from Agassiz. He arranged most of his *Field Sports* subjects in suspended animation and placed them in woods or grasses at eye level, much as Audubon and Wilson had. Elisha Lewis commissioned George G. White and Christian Schussel to make original designs for his *Hints to Sportsmen* under the supervision of ornithologist John Cassin. White, who was drawing insects and birds for the government surveys at the same time, shows an interest in multiple attitudes in his bird compositions much like Audubon's and Wilson's.[12]

Children's literature produced in America had relied on heavily illustrated "books of beasts and birds" since Isaiah Thomas, America's first publisher, issued the Mary Trimmer natural histories adapting Oliver Goldsmith. This type of simplified life history, usually dependent on previously published illustrations, continued through the end of the period under discussion. Images in local productions were quickly copied in cruder engraving from more lavishly illustrated works; for example, the adaptation of a jaguar in a Baltimore children's book reveals a debt to the Parley image (Illustration 5.7).

However simplistic the images or prose seems to modern-day reader/viewers, children avidly read the illustrated books, as their battered appearance reflects. Some of the cheapest works were treasured as mementos and handed down to family members; one inscription in a slim booklet, circa 1835, notes, "Sylvanus Huntoon's

book given him when a boy—I now give it to Hattie to *always* keep—
her mother/ A. R. Clarke./ Feb. 22 1891." Many children drew the
outlines of the animals alongside the illustrations or traced them on
the backs of the pages. One active individual "redrew" not only the
bear but also the lion on the back of the plates of one such picture
book (Illustration 5.8). Thus, youthful readers became intimately familiar with the attitudes and appearance of both foreign and native
wildlife.[13]

The influence of such images, not only viewed frequently but
sometimes copied over and over in childhood, may be incalculable.
Those who devoted their later lives to the pursuit of natural history
linked their careers to the early reading of the textbooks and simpler
illustrated compilations. The taxidermist Martha Maxwell, who later
corresponded with Baird, while a child in rural Pennsylvania, owned
a paperback "Book of Beasts" produced by James Phinney of Cooperstown, New York. Charles Wilkins Webber, an author of western
stories and a friendly Audubon reviewer who himself attempted to
write natural history essays, remembered one book's influence on his
avid boyhood bird study in Kentucky. He believed that he had discov-

Illus. 5.7. "Jaguar." Hand-colored wood engraving by anonymous artist, for *Pictorial Instruction for Girls or Boys: Animals* (Boston: Weir & White, c. 1850), 2. (Courtesy of Winterthur
Library, Printed Book and Periodical Collection.)

184

THE BOOK
OF NATURE
Natural History
in the
United States

ered a new species of mockingbird only to have his older sister correct him by using "a small school edition of selections from ornithology, with wood-cut illustrations . . . she held her hand on the page" to show Webber that his find was the butcher bird. (That edition may have been a Parley production.) Robert Ridgway admitted in recalling how he had pored over his new *Parley's Illustrated Nature* during his Carmel, Illinois, childhood that he watercolored the wood-engraved illustrations (in his words, "the woodcuts [were] disfigured by my efforts to color them"). Frank Chapman, the prominent late nineteenth- and early twentieth-century ornithologist, remembered making "many several remarkable identifications" with his copy of *Illustrated Nature* in the 1860s. The intensity of these reminiscences alone demonstrates the pictures' influence.[14]

The Impact of Dissemination

A host of critics have raised questions relating reader/viewer response, editorial or authorial intent, and the reproduction of images in multiple images and different contexts. The translation of image

Illus. 5.8. Pencil sketch of lion. In John Bigland, *A Natural History of Animals* (Philadelphia: John Grigg, 1828), opp. p. 103. (Courtesy of Winterthur Library, Printed Books and Periodicals.)

in different media and accompanying texts may change informational and aesthetic values. From the ancients onward, natural history illustrations have been copied whenever images of flora and fauna were required, often in ludicrous contexts and translations substantially changing the original. In a well-known example, three prototypes of the rhinoceros wander through three centuries of European art, illustrating the tradition of copying. This period, though significantly shorter, was also one of vigorous imitation.[15]

The translation of high-quality, hand-colored prints into wood engravings naturally changed the images' informational content and visual values. Color, that crucial factor of species identification and aesthetic attractiveness, was lost in the monochrome wood engraving, as was the life-size scale with its dramatic potential and capacity to provide exacting detail. However, wood engravings retained major physical features such as shell markings, toe and claw positions, the number and shape of leaves, and animal shapes, although they omitted finer detail such as tiny feathers around beaks and the segments of insect tarsae. The linear form of reproduction captures the vital outlines and attitudes, as in adaptations of Michaux's plants and Wilson's birds. The influence of Bewick on the Boston engravers enhanced their feel for the relationship of the bird to its background and respect for fine detail. The majority of the Parley, Nuttall, and Harper Brothers images conveyed the general appearance of the originals remarkably well, considering space limitations. As we saw in Illustration 5.4, Audubon's striking design for the wild turkey remains.

The exact retention of details and outlines of the Parley, Harper, and Hall and Bowen engravings was due more to the technologies of stereotyping and electrotyping and practice of selling or renting blocks to other publishers than to a profound commitment by editors and publishers to exactitude (improvements in tracing techniques may have also promoted tighter copying). Before the use of these publishing and technical innovations, wood engravers had to reengrave more frequently and so unconsciously lost or changed forms. Even when engravers redrew blocks, as in the *Ballou's* engravings and in the original Sanborn Tenney work, they paid close attention to markings, texture, and pose.

Other features of this dissemination were the numbers and geographical distribution of the images. The large number of copies sold is astounding compared to the few hundred copies of the original works of Audubon, Wilson, and Holbrook and other scientific monographs. *Graham's* boasted a membership of over 40,000 and *Godey's*

186

THE BOOK
OF NATURE
Natural History
in the
United States

Lady's Book alone had a circulation above 150,000 before the Civil War. Editions of *Naturalist's Library* and the Harper Brothers works must have run to thousands of copies. Textbooks usually were issued several times; Tenney's *Manual of Zoology* ran to at least seven printings. Although historians have emphasized that book distribution favored urban centers, textbooks penetrated the rural Midwest and South, as Ridgway's and Webber's childhood experiences in Illinois and Kentucky and the Macon, Georgia, book of birds suggest.[16]

The persistence of the images is also noteworthy. The Parley images that originated in 1831 reappear up to the 1870s. Images such as Audubon's Baltimore oriole's nest migrated from *Uncle Philip's Conversation* in 1825, to the Harper books on birds in 1839, to *Ballou's* 1856 article on bird architecture—all based on the prototype from the British Society for the Diffusion of Useful Knowledge. The Tenney artists chose to use several of Alexander Wilson's images even though more recent prototypes were available. The illustrations' longevity stemmed in large part from the availability of ready-made images and plates. To create new images for book production required not only capital but great effort on the part of the author or publisher. Elisha Lewis frankly complained to his readers of the unanticipated time necessary to develop illustrations for his second edition of *Hints to Sportsmen,* as he had to obtain specimens for his two artists to draw, persuade the naturalist John Cassin to approve drawings, and then oversee the wood engravers who followed these designs. Tenney's pointed thanks in his introductions to his publishers, Scribners, for the "large outlay" needed for the hundreds of wood engravings subsequently electrotyped underscores that each new drawing or wood engraving required capital. Designs ready to be copied—or better yet, plates ready to be used—cut time and money.[17]

An obvious point often overlooked is the increase in naturalistic, accurate forms in the literature related to natural history outside the major monographs. One need only recall the tiny images pasted in the margins of Webster's *Zoology* or the odd *Massachusetts Magazine* buffalo to realize the tremendous increase in quantity and quality of images in the Parley books and the illustrated magazines. In even the smallest and cheapest illustrations, such as the quickly executed jaguar cut for a children's book, the forms are both lifelike and interesting because of their derivation from the Harvey/Parley prototypes. The new technologies did not increase enormously the number of original designs of the jaguar or leopard, for example, in circulation; instead, they ensured that those anonymous designers wishing to use these types of animals had naturalistic models.

The engravings' availability may have been the key factor for editors and authors in deciding which images and accompanying articles to run. The periodicals' editors favored the wide appeal of interesting and morally pure natural history, but the presence of Nuttall/ Bowen/Hall plates certainly made the illustrated series such as "Wild-Birds of America" possible, just as the extensive British shell imagery encouraged articles such as "Where Ladies' Shells Come From." The Comstock and Parley textbooks could focus on ornithology, thanks to the presence of the wood engravings first produced for Nuttall's *Manual*. Mineralogy and conchology were decidedly more popular activities in antebellum America than ornithology, but because of the Parley and Nuttall images more illustrations of birds were published. This visual presence potently endeared ornithology to the next generation, as exemplified by the young Frank Chapman and Robert Ridgway, transfixed by their illustrated bird books.[18]

Editors' and authors' other consistent contribution to fostering natural history was the retention of image and text. Titles and articles discussing habits and appearance based on the original source usually accompanied the images. The ability to combine wood engraving with print type meant a physical linking of text and image closer than that of lithography or metal engraving. The descriptions of plants' or animals' color, size, and distinctive characteristics encouraged the reader to mentally resize and "colorize" the image to fit the specifications. It is difficult to judge whether all reader/viewers carefully read text and studied images, but the stories of Chapman's remarkable identifications and Webber's recognition of the shrike prove that some were undeniably making the connections. The close physical association of the title to the image was maintained. The constant of a title listing the species' names beneath the image must have differentiated the image from popular animal and flower genres in the reader/viewer's perception and indicated scientific content and accuracy. The editor and reader/viewer assumed that the name, whether in Latin or English form, conformed to the scientifically approved "correct" usage; in the nationwide context of textbooks and magazines, such usage was eliminating the local, vernacular names and promoting uniformity.

The development of mass-produced, steam-driven chromolithography by the 1870s held mixed implications for popular natural history images. Some works featured pages of multiple images of animals and birds derived from Audubon. But generally, chromolithography did not supplant the monochrome line techniques in textbooks and manuals, because it could not print on the text page and had to be

188

The Book
of Nature
Natural History
in the
United States

of the highest quality to render markings and anatomy clearly. One genre, the chromolithographed trade card, did develop natural history imagery for its purpose of attracting potential consumers. Natural history illustration was assumed to kindle interest or at least offend no one. Animal and bird images related to the "living animal" iconography with the accompanying title/identification printed underneath had been a popular subject long before the famous Louis Agassiz Fuertes cards in Arm & Hammer baking soda boxes of the 1920s. Lithographers at Boston's Bufford firm, for example, used Audubon *Birds* images for an extensive Arm & Hammer series of "Beautiful Birds" available for ten cents' postage in the early 1880s. Other trade cards feature birds or animals in the now standard conventions of characteristic pose and suggested backgrounds. The example of the California quail even includes anatomical details (Illustration 5.9).[19]

Natural history forms generally did not permeate conventional flower and plant imagery generated in America between 1825 and 1875. The many popular drawing manuals used examples from the European animal and flower painting traditions, not scientific illustrations, to show beginners how to sketch natural forms (see Illustration 2.7). Designs for decorating mass-produced objects such as sheet music, popular prints, textiles, and pottery similarly relied on the strong flower and animal traditions available in imported publications. Only in specific instances in this "age of flowers" were flowers originally designed for botanical works used. In depicting American wildflowers, authors sought images in botanical works because conventional flower imagery focused mainly on familiar domesticated and imported categories such as roses, violets, lilies, and passionflowers.[20]

The depiction of exotic animals most clearly shows the influence of zoological illustrations. Although the lion and elephant emerged in earlier popular imagery and became stylized in many vernacular traditions, after the 1820s images of many other African and Asian animals were widely available only through images based originally on scientific illustration. Thus, patterns using such exotics as giraffes and jaguars in textiles and theorem drawings are naturalistic in shape and attitude, although the designers likely had little direct knowledge of the animal. For example, professional weavers who designed patterns for nineteenth-century coverlets created animated yet realistic exotic creatures against jungle backgrounds (Illustration 5.10). Although the craftsmen who wove this blanket on Jacquard looms in Ohio around 1850 probably never saw monkeys, giraffes, big cats, or

Be sure and tell your friends who smoke to trade at
PIKE'S,
111 Munroe St., Lynn, Mass.

CALIFORNIA QUAIL
(PARTRIDGE)

Gallinaceous
Birds
F. Grouse.

Illus. 5.9. "California Quail." Chromolithographed trade card. Sold by Pike's (Lynn, Mass.), c. 1875–1882. (Courtesy of Winterthur Library, Joseph Downs Collection of Manuscripts and Printed Ephemera, No. 68 × 164.)

Illus. 5.10. Detail of coverlet by anonymous maker, c. 1850, possibly Ohio. Acc. no. 56.113. (Courtesy of the Metropolitan Museum of Art, Rogers Fund, 1956.)

palm trees, the existence of wood engravings enabled the creation of
identifiable outlines, however different the medium. The crouching
cat perhaps is derived from a British periodical illustration of a
cheetah.[21]

One coverlet, possibly mass produced as a souvenir for the 1876
Centennial, contains a provocative use of a quintessentially native
creature. In the border alongside a resting deer, wild turkeys (derived
from Audubon's memorable model) strut; the many Parley productions containing the turkey wood engraving most likely provided the
direct inspiration (Illustration 5.11). Thus, the coverlet designer had
access to a prized image of a rare work produced fifty years before
because of the widespread dissemination effected through antebellum publishing.[22]

Individual Responses

Historicizing visual response involves as much if not more speculation than does response to text. The provocative, if fictional, example of a young Jane Eyre bypassing the text to prefer the stormy and
dramatic seabird imagery of Bewick's *British Birds* over the charming
scenes of songbirds now favored warns latter-day interpreters of the
dangers involved in inferring individuals' preferences. Moreover,
most viewers, in particular patrons of the arts in this or any period,
rarely described their motivations, likes, and dislikes as tellingly as
Brontë's character or in as much detail as the historian would wish.
Resources specific to this era, such as the voluminous materials associated with Audubon's subscribers, however, suggest which qualities attracted viewers. Therefore, an extended discussion of these patrons'
motivations and responses appears possible. Increasingly popular and
visible decorative arts like wax flowers and fancywork made from natural materials potently imply the importance given to natural objects
in these decades.

Reviewing the Monographs

The natural history monographs did not generate as many published reviews as did the novels of James Fenimore Cooper or Washington Irving, which sold thousands of copies, probably because
periodical editors assumed that few in their audience could own the
scarce works. Just as important, the authors could not afford to send
complimentary copies to many reviewers and so missed opportunities
for press attention. Those magazines that did consistently review
monographs, such as the *North American Review,* promoting its brand

192

THE BOOK
OF NATURE
Natural History
in the
United States

of Whiggish nationalism since 1814, and *Silliman's Journal of the Arts and Sciences,* the most stable U.S. scientific publication of the nineteenth century, noted continually two interwoven themes: the visual elegance of the works and the pride inspired by their national origins. The *North American Review* in 1825 praised Say's *American Entomology* for its handsome typography and the hand-colored plates' execution: "The work . . . affords a most encouraging testimony of the state of

Illus. 5.11. Detail of coverlet by anonymous maker, c. 1876, probably Philadelphia. Acc. no. 1982.243. (Courtesy of the Metropolitan Museum of Art. Gift of Mr. and Mrs. John P. Kauffman, 1982.)

the arts in this country, and as such deserves the patronage not more
of the lovers of science, than of all persons who are disposed to advance the progress of liberal pursuits." Charles Wilkins Webber trum-
peted his and presumably his country's pride in the *Viviparous
Quadrupeds* folio two decades later: "We have at last a great National
Work, originated and completed among us—the authors, artists, and
artisans of which are our own citizens. . . . Let it never be said again,
that American Painters and men of Genius must go to Europe for
engravings and illustrations, because they can find here neither the
enterprise, the means, nor the mechanical skill necessary." *Silliman's
Journal* favored the beautiful copperplate engraving and hand-coloring of the Lawson family in Haldeman's *Freshwater Univalve Mollusca.*
The *North American Review* in a notice of William Starling Sullivant's
1864 *Icones Muscorum,* his work on mosses, admitted, "ordinarily a
special and technical work like this . . . would not call for notice here.
This does so because of its rare perfection as a work of art as well as
of science." The plates are compared to a European counterpart but
are deemed "more exquisite, mainly because they are upon copper
instead of stone, and on the whole are probably unequalled." As late
as 1885, American reviewers continued to judge by the great European hand-colored plate tradition, as reflected in the ultimate praise
given to William H. Edwards's *North American Butterflies:* "They [the
plates] surpass anything that has been given to the world from the
most famed *ateliers* of Europe."[23]

The judgment of later book connoisseurs confirms that these reviewers were not guilty of "puffing" unworthy works, but that the
original natural histories were among the most ambitious and best
executed books, especially in regard to illustrations, in the first six
decades of the century. Encyclopedias, geographies, travel works, and
other genres did not aspire to the same quantity and quality of plates,
with works such as J. T. Bowen's beautifully lithographed imperial
folio *History of the Indian Tribes* (1829–1844) and the smaller books of
the great antebellum illustrators, F. O. C. Darley and John Sartain,
proving the exceptions. The major folios featuring artworks were imported from Europe. The fervid aspirations of the creators to emulate, in the European exile Constantine Rafinesque's words, "the
model of the splendid European publications intended for the
wealthy," set and, in cases such as Edwards's *Butterflies,* Audubon's
Quadrupeds, and Haldeman's *Mollusca,* achieved the goal of matching
their competitors' beautiful and cleanly executed illustrations and
handsome text. The genre's demands for precise detail and subtle
coloring required the finest line reproduction methods and best col-

194

THE BOOK
OF NATURE
Natural History
in the
United States

oring available, in this period generally the traditional copperplate engraving and hand-coloring. However, the exacting color requirements did inspire some innovations in printing in colors: lithotints in Holbrook's *North American Herpetology;* the first chromolithographed book in the United States, William Sharp's *Waterlilies;* and Julius Bien's highly complex chromolithograph version of *Birds of America,* produced in the early 1860s.[24]

Audubon and His Patrons

The works of John James Audubon offer the best opportunities for judging "viewer response" to the original scientific monographs because they attracted the most subscribers and reviews. Extant lists indicate that at least 175 individuals and institutions completed subscriptions for the $1,000 double elephant folio *Birds,* and over 300 subscribed to the imperial folio *Viviparous Quadrupeds* priced at $150. Some 1,050 copies were produced for the octavo *Birds,* and 2,004 subscribers purchased the $30 *Quadrupeds.* Other editions of the *Birds* and *Quadrupeds* under the supervision of the Audubon family were produced with hand-colored plates by Bowen & Company in the 1860s and 1870s. Audubon's celebrity, aided by his striking looks and exciting adventures, and his personal contacts with editors ensured that newspapers would write or copy reviews. Some individual subscribers left evidence of their attitudes toward their purchases in their treatment of books. Their backgrounds and motivations for supporting Audubon provide insights not only to his success but into the nature of other contemporary patronage of natural history and art.[25]

Audubon's patrons composed the new social and economic elite emerging in the decades between 1830 and 1860 and retaining power after the Civil War. Operating mostly out of the urban Northeast, they were responsible for the rapidly expanding economy transforming the nation's transportation and industry. Many invested in the nascent railroad systems while others, including the Grinnell brothers, Thomas Tileston, and Thomas Pym Cope, ran the revolutionary transatlantic clippers. Financiers in the New York firms of Ward & Prime and Alexander Brown & Sons were establishing branches of a nationwide banking system. The famous Massachusetts textile manufacturing families, the Lawrences, Appletons, and Lowells, leather manufacturers such as Charles Leupp, and sugar refiners such as Robert Leighton Stuart and Joseph Lovering were creating and reaping rewards from the "market revolution." "Grocers" Samuel Bispham and the New York partners Luman Reed and Jonathan Sturges built up wholesale distribution networks between the Northeast and Midwest. New Bedford and Nantucket, areas with "new wealth" from whaling, and New Orleans, with its many cotton merchants, contributed octavo *Birds* subscribers. A few, such as Frederic Tudor, the "Ice King," who devised techniques to carve ice from New England ponds and ship it to the Caribbean, and Benjamin Brandreth, who advertised that his patent medicine rejuvenated the blood, established more colorful careers. Some, such as the woodchuck keeper Daniel Wadsworth of Hartford, Connecticut, and the Philadelphian Samuel Breck, inherited wealth. The many subscribing doctors inherited property themselves or married wealthy wives, according to the gossipy booklets popular at midcentury that listed the "five hundred most wealthy" individuals in the major cities. All participated in and profited from an economy far different from that of Alexander Wilson's subscribers.[26]

The Audubon patrons' private interest in natural history and art at times did coincide with public benefit. They generally purchased memberships in local libraries, art galleries, and natural history organizations and supported those institutions' additional projects. They donated natural objects to natural history societies (minerals and shells literally picked up on their travels were especially popular). For example, Robert L. Stuart donated a snowy owl to the New York Lyceum, Daniel Webster gave two oystercatchers to the Boston Society for Natural History, and John A. Lowell's herbarium of over 1,200 flowering plants was donated to the Boston Society. Robert L. Stuart began in the 1840s to build a library of magnificent natural history books, which was later bequeathed to the New York Public Library; he was one of the few American subscribers to John Gould's folios. The majority, who would not have considered themselves naturalists, did not collect as systematically such large collections as Luman Reed's and Robert Gilmor's famous mineral collections, but, like Jonathan Sturges, a partner in Luman Reed's wholesale grocery business, and his wife Mary, they decorated their mantelpieces with shells and collected "beautiful minerals" for their cabinets. Most possessed a general background knowledge of natural history gained through reading as adults, not through childhood education.[27]

The subscribers' art patronage was visible through their frequent loans to art institutions' annual exhibits. Their tastes, as reflected in the loans, shared the prevalent fondness for still lifes, landscapes, genre paintings, and Old Master copies. They gave money to budding national organizations such as the National Academy of Design and the American Academy of Fine Arts. A majority of the most distinguished collectors of the era—Luman Reed, Thomas H. Perkins, Rob-

196

THE BOOK
OF NATURE
Natural History
in the
United States

ert Gilmor, Reed's partner Jonathan Sturges, fellow New Yorker Charles Leupp, South Carolinian Robert W. Gibbes, James Robb of New Orleans, and Daniel Wadsworth—subscribed to one or more Audubon works. The other subscribers did not plaster their walls with art, as Perkins was wont to do; nor did they possess twenty-four folios of European prints, as Robert Gilmor did. Instead, they bought prints, paintings, and sculptures from American artists and European sources to decorate their homes, particularly the parlors and libraries. Some owned folio books of prints reproducing the artworks in the Louvre and Dresden Galleries, which they would pore over in the evenings or Sunday afternoons, "calling attention to their endless beauties," in the words of one contemporary.[28]

It was no coincidence that these men, so attuned personally and publicly to art, science, and literature, patronized Audubon. He and his supporters knew of and sought out men well known for these tendencies, much as Alexander Wilson had scoured streets in Charleston for wealthy homes and relied on lists from the friendly librarian. Audubon's early success in Great Britain was partially due to his Liverpool friends' generously supplying letters of introduction and support to likely patrons in Edinburgh, Manchester, and London, who in turn introduced him to friends in their cities. He refined this technique of personal introduction and display of sample plates in extended trips to America during the production of the folio *Birds* and so built up a reservoir of support. Upon the folio's completion in 1839, he decided to live in New York City, rather than England or other U.S. cities, because he recognized it as the center of art and literature and wished to draw upon its burgeoning population's purse. His continuing efforts to publicize his works and the connections he built through the *Birds* created a synergy that won thousands of octavo subscribers and enabled him and his sons to produce the folio *Quadrupeds* plates from 1842 to 1848.

Several factors in the antebellum United States aided his progress, such as newspaper editors eagerly writing and copying articles about his personality and productions to fill their columns. Previous purchasers could easily identify other likely subscribers because, as members of the same socioeconomic elite often linked by family or business ties, they attended the same art exhibitions, joined the same organizations, and knew the tastes and wealth demonstrated at parties and through home furnishings. The relatively limited leisure activities in this era of religious moralism included natural history, art, and literature—the components of Audubon's work. Often subscribers themselves not only wrote letters of introduction but personally escorted the naturalist to their friends' homes and took care of business matters for him. After the elder Audubon's death in 1851, Victor Gifford Audubon and John Bachman carried on the tradition of personal solicitation, with Victor garnering hundreds of "signatures" on his 1853 quest for octavo *Quadrupeds* subscribers in the South, visiting South Carolina, Alabama, and Louisiana, and Bachman stirring his Charleston compatriots to contribute "four hundred good and true names."[29]

Primed with this background, one may hypothesize about the subscribers' reactions and motivations. Their own treatment of the volumes may act as additional potent evidence. In their homes, considered the most elegant and tasteful of their day by contemporaries, they prominently displayed the works in the most public spaces, the parlor and library. The Philadelphia banker James Dundas's copy of the *Quadrupeds* folio rested on a portfolio rack in the center of his well-furnished library. New York banker Robert Ray laid his newly received first volume of the folio *Quadrupeds* on his library table so that guests at a soiree could gaze at it. Luman Reed placed his *Birds* in his art gallery open to the public, where Cole's paintings and Reed's cabinets of minerals and shells also resided. The Philadelphia doctor Charles Frederick Beck kept his folio *Quadrupeds* in his parlor with copies of the *Journal of the Academy of Natural Sciences,* four issues of the *Art Journal,* and a French folio of engraved art reproductions. Years after the event, Mary Sturges, whose family patronized Audubon, recalled first seeing the folio *Birds* on a "newspaper editor's" central parlor table. Owners lavished special treatment on the works, such as customized bindings. Daniel Webster did not keep up his payments but did, however, spend the money to commission a special mahogany cabinet especially for the *Birds.* The New York City Fire Department might hold the honor of ordering the most elaborate bookcase for an Audubon work. Their splendid Renaissance Revival bookcase made by Brooks Cabinet Warehouse housed a copy of the octavo *Birds* presented to the acclaimed singer Jenny Lind on September 13, 1850.[30]

Such pride in possession and display of taste and wealth did not preclude an intense visual experience. Today's viewers may share the subscribers' awe of the plates' execution but must perform an imaginative stretch to appreciate the impact that the mimetic color images made in the century before color photography and moving pictures. Comments suggest that onlookers pored over details and eagerly viewed plate after plate. George Templeton Strong noted in his diary that he and a group looked at "Audubon's foxes and squirrels and

198

THE BOOK
OF NATURE
Natural History
in the
United States

skunks for an hour" during a party at Robert Ray's mansion. The Library of Congress copy of the *Birds* proved so popular that its pages quickly revealed its frequent use. William Oakes looked at a volume at the Boston Athenaeum for over two hours; after faulting the quality of the hand-coloring on the bills (the neatness of copies was a major concern to subscribers), he softened his verbal blow by gushing, "the wonder is still to me, how so many acres of our forests and their inhabitants can be afforded so cheap."[31]

Aspects of interior decoration further suggest that textures and natural objects of many types fascinated these viewers. As a later section will discuss, upper- and middle-class households decorated their homes with such real natural objects as shells, stuffed birds, and grasses, and increasingly women incorporated these same natural materials into their handicrafts. In the words of one cleric who subscribed to Audubon, "The texture of a leaf, or the tinting of a shell, [is] another inlet into the workmanship with which the mysterious universe teems with such continual demonstration of an ever-present God." Thus, these viewers ardently appreciated a wide variety of natural textures on a daily basis and were familiar with details such as bird claws and mineral shapes today considered less than appealing. Little wonder that contemporaries studied every feather, claw, and branch in the prints.[32]

The success of the printing techniques and coloring in recreating the textures and overall quality impressed the original subscribers, most of whom were not well versed in the great European natural history folio tradition. Unlike Audubon's English and French subscribers, they had little experience with contemporary European high-quality, large-format works such as Prideaux Selby's *Illustrations of British Ornithology* and the botanical monographs of Pierre Redouté because libraries and private individuals in America did not subscribe to them. The contrast between the Audubon plates and the small, often uncolored natural history illustrations in encyclopedias and older works such as Wilson's *Ornithology* and Say's *Entomology* must have startled the viewer/owners. Joseph A. Moore of Greensboro, Alabama, wrote Audubon that his copy of the folio *Quadrupeds* "surpasses in correctness of representation and elegance of finish and production of the kind that I have ever had the pleasure of examining."[33]

A comparison with other artistic genres popular among the subscribers and their milieu offers historical insight to their visual response. The viewer/patrons may have appreciated aspects common to other genres they owned or with which they were familiar. Still life's concentration on textures and forms, especially in flower and fruit subjects, relates to the Audubon plates' blossom and branch accessories, frequently praised in reviews. Since several patrons, including Luman Reed, Charles Leupp, and Daniel Wadsworth, bought works from the emerging New York school of landscape painting, and others listed landscapes in their collections, the vistas of American meadows, prairies, deserts, and cities must have interested them. Victor Gifford Audubon, who specialized in landscapes in his own artistic productions, provided lavish backdrops ranging from pastoral river views to unspecified western landscapes in the *Quadrupeds*[34] (see Illustration 2.2).

Genre painting and prints based on imagery of recognizable types performing characteristic actions surely enhanced viewers' reception of these plates depicting "life histories," similarly dependent on representative specimens in typical poses. Jonathan Sturges, for example, owned William Sidney Mount's *Farmer's Nooning*, featuring the stereotypic characters of lazy blacks and wholesome farm boys. Other popular antebellum genre scenes featured mother and child, the happy family, or a solitary mountain man astride a horse. Charles Wilkins Webber's telling description of the opening plates in the *Quadrupeds* in a *Southern Review* article suggests the close link of Audubon's representation of a species in its typical setting to the generalizing inherent in genre painting:

> His plates are true Pictures in the highest sense. Each one of them is complete in itself and tells a story not to be mistaken. They strike one as unitary fragments from the memory of his long life of wanderings, reproduced complete in all their parts—not alone the creature itself in some striking attitude, characteristic of its habits, but, as well, the very scene in which it was first observed. . . . [A] group of Elks standing and lying beneath the shadows of the bordering trees, is seen looking out upon the undulations of those vast prairies of the Upper Missouri. . . . So the fatal eye of the Canada Lynx, with the yellow heat of ferocity in it, compels a sort of shiver from us as we see it in the act of springing upon its unconscious prey, amidst the broken rocks, the decaying logs and tangled firs of a Northern forest.

Patrons would have understood and appreciated not only the high drama of the mockingbird/snake confrontations and the dramatic birds-of-prey scenes but also the majority of plates featuring birds and

200

THE BOOK
OF NATURE
Natural History
in the
United States

mammals pursuing food, feeding young, or verbally communicating via growl or song.[35]

Gender and location held mixed implications for viewing. Few women actually bought the Audubon works, but Audubon claimed that women's wholehearted support often persuaded male family members. Women and men viewed the works together on social occasions or at home. Glimpses of the historical response indicate that the same values were appreciated. Bostonian physician George C. Shattuck, who was also a personal friend of Audubon, and his niece enunciated their shared belief that Audubon the naturalist was "God's interpreter": Shattuck wrote to Audubon of his niece, "She reserves her greatest respect for the interpreter of God, in other words the naturalist." Mrs. Barnwell Rhett, after reading her husband's copy of the octavo *Birds,* told her sons tales of Audubon, the brave naturalist. Her husband, a congressman from Georgia, proudly explained this transmission of values: "Mrs. Rhett takes real interest in your labours, and often describes to my little boys, in glowing terms, taken from your works, of the toils and the pleasures—the labours and the glory of being a great & enthusiastic Naturalist like Mr. Audubon." Elizabeth Fothergill, an old friend from Liverpool, complimented Audubon on the successful reduction of the mockingbirds in the octavo edition. Avid hunters such as Daniel Webster might have argued that women lacked "field experience" to judge wildlife in its environs, yet neither men nor women were familiar with many species in any case. Audubon concentrated his efforts on cities, so his subscription lists favored the Northeast. Subscribers from the Midwest and more rural South did manage to hear of and subscribe to the works, however, as did that fervid admirer, Joseph A. Moore of Greensboro, Alabama. Victor Gifford certainly found southern cities like Savannah and Charleston receptive to his octavo *Quadrupeds* overtures, to judge from the lavish number of subscriptions. Nationwide distribution adds credence to the previous suggestions that natural history was a unifying element among the literate classes before the Civil War.[36]

Art historians provocatively have linked this same group of individual patrons including Jonathan Sturges, Charles Leupp, Luman Reed, and Daniel Wadsworth, to attitudes expressed in genre and landscape painting that they commissioned and owned. Through its use of sexual and racial types, genre painting may have given expression to hidden frustrations in its viewers and reinforced the dominant-class superiority. The favored panoramic viewpoint of overlooking the landscape may have embodied human control over nature and rationalized the new economy's rapid development of natural resources. These same men's support of Audubon opens the issues of power and justification as the depicted birds and animals disappeared along with pristine landscape as a result of economic activities linked to financial interests. The perfect replication of nature may represent the ultimate domination over nature as in the panoramic landscape.[37]

More likely, viewers entered, to some degree, the "ornithocentric" or animal/bird viewpoint implied by the plates. To continue the psychological metaphor of the other, the "other-object" draws enough interest to become the "other-subject." Readers' natural history interests imply that they gave weight and consideration to the "differentness" of birds and animals. Daniel Wadsworth, the noted art patron, himself placed his hibernating pet woodchuck in the parlor to observe whether it would awaken. The appreciation of similarities mingled with this awareness, for the subscribers enjoyed anecdotes of animal intelligence and character, and in a few cases expressed the rising humanitarian movement's belief that animals deserved people's consideration. How could they explain the destruction of the creatures whose lifelike representations "lived" in their homes? Because they took to enjoying natural history as a refuge from their business lives, not surprisingly they appear to have disconnected their economic activities from this leisure pursuit. If asked, they probably would have cited the argument prevalent in their milieu, that the creatures as well as landscape were doomed and that at least these facsimiles would remain. The aldermen of the City of New York thus explained their rationale for buying the *Birds:* "These lovely tenants of the wood, can only be *thus* perpetuated; for many of them will disappear with the forests which are falling before our advancing population." The process of representing nature, therefore, did not inspire immediate reaction against the economic structure but only palliated its outcomes.[38]

The uniqueness of Audubon's success now may be understood better. There was a large enough audience to support natural history works potentially in a variety of disciplines, including conchology and botany. The building of several observatories and funding of other science books, such as Louis Agassiz's 1861 *Contributions to the Natural History of the United States,* featuring plates of invertebrates, which drew some 2,500 subscribers, further indicate this potential. This audience, however, required powerful and direct persuasion, preferably provided through personal contact. Audubon and Wilson tirelessly provided this "ocular inspection," but other naturalists, such as

202

THE BOOK
OF NATURE
Natural History
in the
United States

Thomas Say, S. S. Haldeman, Daniel Giraud Elliot, Asa Gray, and William H. Edwards, were content to send out prospectuses and solicit only local contacts and professional colleagues. Their subscription lists are correspondingly weak. As the Audubon scholar Ron Tyler has demonstrated, the octavo editions did give the Audubon family their livelihood, but their productions represented an enormous investment of time and dedication that the other naturalists understandably could not make. The constant demands of travel would have consumed time needed for writing and supervising plates. Moreover, most had other livings to earn and other ways to make their reputation. Gray and Agassiz, for example, had many other conduits for their expertise, so they could abandon their lavishly illustrated projects (*Contributions* ended after a few numbers); Audubon and his English counterpart John Gould, in contrast, had staked all their careers on their own publications and so were compelled to maintain productions despite discouragements.[39]

The changes in audience interests after the Civil War marked the end of American natural history illustration on the grand scale. Although publications involving wildlife, such as the essays of John Burroughs and Elliott Coues's *Key to North American Birds,* proved immensely popular in the 1870s, the rare union of art, nature, and science was fragmenting. The motivation of national pride, so clearly expressed in the published reviews, waned. The acquisition of foreign objects satiated national pride, as wealthy Americans interested in art turned to European academicians rather than native artists, and to human artifice rather than nature. Audubon subscribers August Belmont, Alexander Turney Stewart, and Francis Calley Gray accumulated extensive collections of Chinese porcelain, French paintings, and European reproductive engravings, respectively, in the decades after their Audubon purchases. The English audience, by comparison, supported its well-entrenched natural history folio tradition well into the twentieth century. Daniel Giraud Elliot's subscription lists for his magnificent folio-sized works after *Unfigured Birds of the United States* indicate British support, and he, in fact, lived in England for ten years. The elaborately illustrated natural history monograph, the format inspiring Wilson and the two succeeding generations, proved short-lived yet influential in the United States.

Using the Conventions

Natural history illustration undeniably "looked" different from other representations of plants and animals. Its conventions implied that artist and viewer held the naturalist's peculiar focus on the species representative, free from distractions of lighting and atmosphere, with the landscape subordinate. Specialized as the genre was, Americans of this period not generally associated with natural history not only knew but employed the visual part of the discourse to express their relationship to the nonhuman world. Those travelers who were previously acquainted with the genre could use it to make sense of new circumstances in much the same manner in which those who wrote about experiences used the verbal discourse. Some Americans celebrating the beauty and moral significance of "Nature" in their artworks and handicrafts demonstrated knowledge of natural history illustration.

One reason why those not associated with natural history could employ visual discourse was that it was usually self-taught. Most natural history artists learned largely by copying natural history illustration, for no separate academic course or drawing manuals geared to the subject existed. Maria Martin, the Charleston, South Carolina, resident whose drawings were published in Audubon's *Birds* and Holbrook's *Herpetology*, for example, honed her skills in executing butterflies for Audubon by copying Thomas Say's *Entomology*. The young Spencer Fullerton Baird copied outlines of animal heads from textbooks and was privileged to use plates from the folio *Birds* in Audubon's own studio. Perhaps the most inveterate copyist who exercised his skills over the longest period of time was the Bluehill, Maine, parson Jonathan Fisher, who was self-taught in landscape and portraiture as well as natural history. While a student at Harvard in the 1790s, he copied not only George Edwards's images of exotics in foreign lands in *Gleanings of Natural History* but also its columns of French and English. Notebooks show he copied pages of shells, plants, and Bewick's *Quadrupeds* which he used some four decades later for his 1832 *Scripture Animals*.[40]

Well-educated men and women who drew and painted thus felt no qualms in experimenting in the genre. Samuel Breck, an Audubon subscriber from Philadelphia, for instance, is not unusual when he speaks of copying a bluebird and wren, perhaps from his editions of Goldsmith or Buffon, into his journals. Victorine, Sophie, and Eleuthera, the daughters of Eleuthère Irénée du Pont, who himself studied under French botanists and aided Michaux, developed great skill in copying plants from the many botanical volumes in their father's library. Sophie especially excelled in entomology, mimicking Donovan's and Say's plates in illustrations intended for the "Entomological Society of the Brandywine," a lighthearted group formed by the du Pont children and their friends to celebrate their natural history study.[41]

204

THE BOOK
OF NATURE
Natural History
in the
United States

Armed with the ability to produce in the genre, acquired only after many hours of painstaking practice, these individuals drew the flowers from their own encounters in the highly conventionalized forms. Breck and the du Pont sisters drew flowers fresh from the gardens and fields. Sophie recorded the life histories of unfamiliar caterpillars by watercolor sketches of their metamorphoses. Fisher drew the native flowers and, in one instance, an unusually shaped carrot with multiple roots from his garden, which merited a permanent record in his opinion; he proudly labeled these efforts "from Nature."[42]

On the Homefront

Natural history's visual conventions permeated certain compositions of art in oil and watercolor and handicrafts not directly linked to the disciplines. A few academic artists in these decades, such as Arthur Fitzwilliam Tait and William J. Hayes, who specialized in animal art, did use Audubonesque forms to depict deer and game birds with chicks. However, uses of natural history iconography more dramatic and obvious exist in a few works of nonacademic or little-known artists. The oil painting *Birds,* executed by an unknown artist perhaps in the 1840s, unfortunately provides little information about its artist (Illustration 5.12). Fourteen different species rest on branches or, in the case of the meadowlark, crouch on the ground. The similarities to the finest ornithological illustration are marked. The artist chooses species, such as the bobolink, red-tailed hawk, and catbird, not usual in conventional imagery; he or she skillfully handles the difficult outlines of the primary and secondary feathers and distinguishes consistently defining marks such as the flicker's cheek patch. The poses reveal a Wilsonian or Audubonian animation, and, in fact, the eagle is composed after Audubon's "Bird of Washington," the meadowlark perhaps after Wilson's version, and the nighthawk probably from Thomas Bewick. The artist probably had access to the wood-engraved derivations and used the color descriptions in texts and actual specimens to supplement his or her own visual observations. Unfortunately, lack of information about the painting's creation makes the motivation uncertain, but the portrayal testifies to the painter's knowledge of living birds and to the inspiration of printed sources.[43]

Commonplace books and friendship albums—those compilations of sayings, poems, and illustrations from family and acquaintances that abounded in this era exuding sentiment and love—often contained watercolors of flowers and natural objects because of the frequently enunciated link between nature, beauty, and godliness. Flowers in particular were directly associated with love and affection.

Illus. 5.12. *Birds,* by anonymous artist, c. 1840. Oil on canvas, 17 × 14 inches. (Gift of Edgar William and Bernice Chrysler Garbisch. © 1997 Board of Trustees, National Gallery of Art, Washington, D.C.)

206

THE BOOK
OF NATURE
Natural History
in the
United States

Most of these paintings were drawn from print sources ultimately derived from the flower-painting tradition (bouquets or nosegays of a variety of flowers, for example), but a few cases show the direct influence of botanical illustrations. Between 1816 and 1839, Susan Nichols of Fairfield, Connecticut, bound into her book watercolors of native berries and fruits, drawn by herself and perhaps others, evidently valuing them as keepsakes. The image of the "moose bush" shows hesitancy in outline and paint handling but a clear knowledge of botanical illustration principles, including full presentation of the leaf and fruit and the titling of the species (Illustration 5.13). (The 1838 Charleston, South Carolina, album of Eliza Bachman holds a bird drawn by a budding ornithologist, John Woodhouse Audubon.) Eleuthera du Pont's album coupled images of flower bouquets with poems selected by friends and relatives. Her sisters and friends contributed scientifically oriented images as well, such as Sophie's butterfly and trumpet honeysuckle complete with hairy root (Illustration 5.14) and a moth cut out of another sketchbook. These drawings merited their special places in the albums, because they so attentively represented nature and demonstrated as well the affection of the maker.[44]

From the 1820s to the end of the century, upper- and middle-class homes across the United States contained growing numbers of both natural materials and replicas of natural objects. Women were decorating not only with single shells on their mantelpieces but with "sailor's valentines," compositions of shells pasted into stylized designs. Ladies' magazines showed their readers how to arrange feathers and grasses into wall mounts, make paper flowers, and "draw birds with feathers." In the last, the creator pasted feathers from the dead specimen onto the outline drawn on paper, adding beak and legs in watercolor. In making wax flowers, the makers dissected real plants' parts for models of the individual leaves, petals, stamens, and other components; after cutting the wax and wiring the structure, the creator would arrange the flowers into a bouquet and place it under glass. The popularity of these decorations and the substantial time spent in their construction evidenced an intense appreciation for the variety of natural textures and forms.[45]

Taxidermists appeared to have constructed and sold "decorative taxidermy" creations since at least the 1840s, with the practice peaking later in the century. Decorative, as opposed to "scientific," taxidermy composed multiple stuffed birds in naturalistic poses on real or papier-mâché branches set into glass boxes lined with moss. (Naturalists' specimens, if stuffed, were placed singly on simple stands.)

Illus. 5.13. "Moose Bush." Watercolor drawing, probably by Susan Nichols, from Nichols, "Diary." (Courtesy of the Winterthur Library, Joseph Downs Collection of Manuscripts and Printed Ephemera, No. 65 × 607.)

Le jeune papillon, echappé du tombeau,
Qui sur les fruits naissans, qui sur les fleurs nouvelles,
S'envole frais, brillant, épanoui comme elles,
Jouit moins au sortir de sa triste prison,
Que le sage au retour de la belle saison.

May 30ᵗʰ 1823.   Sophia Du Pont.

Illus. 5.14. Swallowtail butterfly and trumpet honeysuckle(?). Watercolor drawing, probably by Sophia du Pont, 1823, for Smith, "Album and Scrapbook," 9. (Courtesy of the Winterthur Library, Joseph Downs Collection of Manuscripts and Printed Ephemera, No. 65 × 623.1.)

Thure Kumlien, the Swedish naturalist and farmer, traded his boxes of stuffed songbirds for other goods in rural Wisconsin, including his eventual copy of *Wilson's Ornithology*. Single stuffed owls were favorite ornaments for parlors and libraries because of their imposing appearance and association with learning and wisdom. Taxidermists, often hunters and "field" naturalists in their own right, knew their subjects' habits and characteristic poses and probably avidly studied the ornithological literature. They could invest the birds with animation and give the minimal yet realistic background that made their groupings reminiscent of the contemporary "living animal" natural history vignette. Joseph Batty's guide to decorative taxidermy, in fact, provided illustrations in the natural history mode for the student to emulate (Illustration 5.15). The owners of these creations may well have been fascinated with the lifelike attitudes, not so much resembling the flitting glances of everyday experience as the fixed proximity of natural history.[46]

Needlecrafters adapted natural history forms for a type of quilt widely identified with the mid-nineteenth century, the album quilt. Experts are constantly bringing to light examples of how women, rural and urban, southern and northern, commented on issues meaningful to them and their communities through their choice of figurative designs. These vehicles of personal expression, composed of individual squares sometimes numbering more than one hundred

Illus. 5.15. "Pin-tail Duck Properly Mounted." Wood engraving by "E. F." for Batty, *Practical Taxidermy*, 104. (Courtesy of the Winterthur Library, Printed Book and Periodical Collection.)

210

THE BOOK
OF NATURE
Natural History
in the
United States

blocks, often were deemed showpieces, given to others on special occasions and displayed at exhibitions. Because women as well as men highly valued nature's role as a source of beauty and moral significance, not surprisingly many quilts feature flora and fauna. Although most floral and animal designs are stylized or drawn from flower painting images, a few show the designer's familiarity with natural history illustration (see Illustration 5.1). The Benoni Pierce quilt square designer, Julia Burdick, whose work opened our chapter, could have seen the stag from Bewick's *Quadrupeds* in many editions of the Parley books or the children's "books of beasts" using wood engravings that were issued in New York state. (These wood engravings reverse Bewick's designs as does Burdick's work). She creates her own striking interpretation of the animal's form using patterned fabrics and gives it a memorable facial expression. However, the elegance of the figure as seen in nature and portrayed by Bewick remains.[47]

Some individual quilters, most based in New York state in the mid-nineteenth century, designed entire quilts on a botanical or zoological theme, with each square holding a separate species. Often the creator titled the square, a practice apparently rare with other album quilts but standard in natural history illustration. Native and foreign species often included in children's and adult's natural histories from the 1820s onward populate these quilts with identifiable outlines, further indicating that the designers had knowledge of realistic representations. Designers of the album quilt form compared squares to individual pages of an album; in this example, the squares relate to the series of illustrations in popular natural histories.[48]

Chloe Barnum Kimball, who had moved from New York state to Michigan by 1853, created in 1869 such a quilt reminiscent of the format she may have known in her home state (Illustration 5.16). The work she entitled "An Ornithological Quilt" displays many species including (in her words) "The Lark," "The Lyre Bird," and "The Baltimore Bird." The lively profiles and plant accompaniments relate to ornithological illustration, yet her many unique designs manifest an individual celebration of the natural world's variety. This labor of nature love was shown not only in the home but outside it: it reportedly won a major state fair prize.[49]

The Travelers

Those encountering totally new landscapes also recruited drawing as a tool to order their often chaotic sense impressions. As the previous chapter showed, individuals of this period were on the move from Europe to America or in America to new homes. Drawing, like writing, could acclimate the artists to their surroundings. Some were required to illustrate their new environs as part of their job, while others graphically strove only to satisfy themselves and their intimates.

The artists who accompanied the federal government expeditions were not only organizing their own perceptions but ostensibly demonstrating to the nation its recent acquisitions. They were acquainting those in the East in absentia with the new natural productions, making them understand a new part of their shared nation. John Mix Stanley, Richard and Edward Kern, H. B. Möllhausen, Arthur Schott, and Robert Ridgway were among the artists whose field drawings were translated into prints in final reports. Some of these

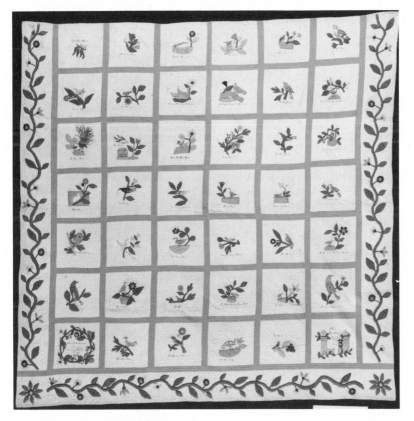

Illus. 5.16. "An Ornithological Quilt." Chloe Barnum Kimball, 1869, Bennington, Mich. (Courtesy of the Michigan State University Museum. © 1987 Peter Glendinning, used by permission.)

Illus. 5.17. Sketches including squirrels, from Titian Ramsay Peale, "Sketchbook," c. June 1820. (Courtesy of the Yale University Art Gallery. Gift of Ramsay MacMullen, M. A. H., 1967.)

M

Rosa

Jamy

Ca

Felis Catus
with two eyes

214

The Book
of Nature
Natural History
in the
United States

artists sketched not only natural history but also topographical views of scenery; in both cases, they ordered the new views to the conventions inherent in both genres. The naturalists could make only relatively rough outlines of mammals and plants "in the field" on the expeditions and mark down measurements, reserving execution of the necessary details—in particular making the multiple anatomical cross-sections—in more controlled circumstances. John Mix Stanley, in his "Memoranda in Relation to Sketches in Natural History, Geology, Botany, and to Views of Scenery and Natural Objects," published in the *Pacific Railroad Expedition Reports,* notes what to sketch: "All that is necessary in regard to the delineation of animals, birds, and reptiles, while on a journey, will be to make sketches of their attitudes and outlines, without going into any minute detail. This is less necessary for birds than for other mentioned reptiles, as frogs, toads, and salamanders, which ought always to be sketched while alive. Details can always best be supplied in the office. . . . Reptiles, fishes, crustaceans, and soft animals generally . . . fade more or less, and should have sketches of color made while alive, or immediately after death. These can be made on the outlines of the natural attitude, but no care need be bestowed in details of drawings, as these will be made anew." The "outlines of the natural attitude" prove essential in the iconography of the "living animal." Because non–field artists such as John H. Richard working in Philadelphia and Cambridge, Massachusetts, also adhered to these same conventions and imparted a sense of life to the coiling reptiles, lively amphibians, and stepping foxes, their images resemble those of the field practitioners.[50]

Some artists were so fully versed in natural history conventions that even the hastiest sketches probably not intended for publication demonstrate these poses. Titian Ramsey Peale, the scion of the famous artist-naturalist Charles Willson Peale, while acting as the artist for the Long Expedition, captured his impressions in notebooks with little time and in far from ideal conditions. The sketchiness of the squirrels on a rock betrays its quick rendition (left portion of Illustration 5.17). The lively sideview and frontal attitudes with rock accompaniment, however, relate to the natural history illustration convention of the living animal in its suggested environs. The schemata thus organized Peale's perceptions about a new subject perhaps observed briefly in hitherto "foreign" territory.[51]

Americans outside the scientific expeditions and publications also drew while in transition, although for more private purposes. Those making memoirs of their journeys to share with family and associates created their own natural history illustrations to enhance

Illus. 5.18. Watercolor drawing, including "Service Berry Tree" and "Hazel-Root," from Lewis H. Miller sketchbook, "Guide to Central Park," 1865. (From the Collections of the Henry Ford Museum and Greenfield Village.)

216

THE BOOK
OF NATURE
Natural History
in the
United States

their narration. Lewis H. Miller (1796–1882), a Pennsylvania German carpenter who made hundreds of pictures in York, Pennsylvania, and later in Virginia, lovingly recorded his European and American journeys. Using a vibrantly individualistic manner in watercoloring, he drew and colored elaborate travel journals from memory, aided probably by sketches taken on the spot. His 1865 "Guide to Central Park" depicts its sites and dells with his adjoining text lending biblical overtones to this "wilderness" and "place of rest." To give a fuller description of the landscape, he lists the park's botanical bounty of flowers and trees in German and English terminology in his conclusion (Illustration 5.18). Evidently knowing the visual conventions of botany through the Pennsylvania German herbal tradition, he drew specimens of hazel root "flower and leaves" with its root exposed and the fruit of the "service berry" above his rendition of park scenery.[52]

The life's work of one mother and daughter demonstrate that those equipped with the conventions could use them to provide psychic comfort in adapting to new surroundings. Orra White Hitchcock studied drawing, botany, and astronomy in her early life in Boston and western Massachusetts before teaching the same subjects at Deerfield Academy beginning in 1813. Her marriage to the clergyman and geologist Edward Hitchcock gave impetus to her artistic and scientific powers, as she drew landscapes, maps, and fossil specimens for his numerous publications. Apparently for her own pleasure she made up an album of her botanical drawings, which she called "a parvum herbarium." When she moved from her secure and popular position in Deerfield to new surroundings in New Hampshire in 1821, she drew sketches of the local mushrooms compiled for another book of sketches, titled *Fungi picti*. Perhaps the familiar occupation comforted her during a difficult time of transition and made her new environment more stimulating.[53]

She must have taught botanical drawing to her daughter, Emily Hitchcock Terry (1838–1921), as the latter was drawing flowers "from nature" as a twelve-year-old in 1850. Lessons at Cooper Union in New York City further refined her skill in watercolor. While taking care of her consumptive husband, Terry took to the fields around her new home in Minnesota and in various southern states where she and her husband sojourned during his recuperation in the 1870s. The resulting watercolor drawings of flowers and sedges exemplify patient labor in trying circumstances, as she took advantage of her nomadic life to explore new areas (Illustration 5.19). Terry later gathered 142 flower paintings from ten states into an unpublished "book," as her mother had. Its title, "American Flowers," indicates her own wide

Illus. 5.19. *"Arethusa bulbosa."* Watercolor drawing by Emily Hitchcock Terry, 1877, Minn., later placed in her "American Flowers" compilation. (By permission of Smith College.)

218

THE BOOK
OF NATURE
Natural History
in the
United States

experience of the continent from the Rockies to Florida. It also poignantly mirrors the national coverage of the published monographs.[54]

National history monographs thus influenced personal ambitions and means of expression. Distinct images such as the Bewickian stag and the Audubonian turkey and the schemata of natural history illustration in general became widely available, thanks in large part to the wood engraving boomlet within the print explosion. Those few Americans who patronized the natural history monographs enabled creation of original images, recycled in some cases many times over. Perhaps Audubon, Wilson, and Asa Gray never imagined their illustrations appearing in such mass-produced products as periodicals, textiles, and trade cards, let alone in personal creations such as oil paintings. Certainly the important works of what would become the life sciences never exerted such impact before or since.

The imagery related to the American monograph specifically became part of the dissemination of natural history. The animated features in characteristic profile against a suggested background, animal and plant, survived the wood engraving process and reappeared both in figures copied from illustrations and in sketches drawn from life. Such images carried connotations of natural goodness and thus were appropriate to grace the home as keepsakes and decorations. This iconography continues but in different genres.

6

Conclusion:
The Figure in the Landscape

❧ *The books are only the guide, the invitation.*

John Burroughs discussing bird manuals in
Wake-Robin, first published in 1871[1]

*I*llustrated natural history books did not vanish from the American scene after 1875, although the elaborate hand-colored monograph tradition faded. In fact, ornithology and entomology as hobbies thrive today owing in some measure to the popularity of illustrated books. The peculiar focus assumed in the period between 1825 and 1875, moreover, continued, for the living form in its characteristic pose and landscape has existed in various reconfigurations until today. In mass media, specialized publications, and conservation legislature, the individual interpreted as representative of a species is an essential component in understanding the natural world.

The fifty-year period we have examined ends with a new phase of the transportation and commercial revolution. The increase in rail-

220

THE BOOK
OF NATURE
Natural History
in the
United States

road lines and rebuilding of industries created a post–Civil War wave of communication and institutional blossoming. Public schools and universities were chartered, and cheap mail routes promoted another boom of periodical publishing. Opportunities for leisure activities diversified. Specialized organizations occurred across religious and scientific spectra. These developments affected the discourse of natural history.[2]

The State of Natural History in 1875

In 1875 the Smithsonian Institution sent out circulars to its correspondents requesting their current address, occupation, and "subjects in which especially interested," and asking, "Have you any private collections, and if so, what?" The collections in question could "include specimens of Mineralogy, Natural History, Ethnology, etc." Since 1850 the secretary of the Smithsonian, Joseph Henry, and his associate, Spencer Fullerton Baird, had built up a network of informants such as P. R. Hoy and the Reverend Charles Mann. They not only contributed natural history objects (as discussed in chapter 3) but also recorded weather data. The number of complaints regarding failures to deliver the *Annual Report* to homes indicates that these correspondents remained keenly interested in the organization's aims and accomplishments. Although the primary goal of the circular drive was to bring the mailing list up to date, particularly concerning deaths of correspondents, Baird unwittingly provided insights into the past and future of natural history study, with implications for the role of reading materials.[3]

What immediately strikes the modern readers of the some 650 circulars is the very fulfillment of the grandiose beliefs expressed in the rhetoric promoting natural history in the previous decades, albeit for a limited number of Americans. Almira Lincoln Phelps, the botanical textbook author, had earlier confidently questioned: "[A]lthough Europe may boast of many stars which irradiate her firmament of letters, . . . may we not justly feel a national pride in that more *general diffusion of intellectual light, which is radiating from every part, and to every part of the American republic!* [my emphasis]." Despite Phelps's assertion (and her hefty book sales), not every individual in the American republic was equally exposed to the effulgence of natural history; and of those exposed, even fewer studied the sciences. Only literate Americans reading and viewing current literature in the decades before 1875 could easily become familiar with the natural history discourse in a variety of genres. Still, many were able to employ the discourse for specific purposes like describing new creatures and creating keepsakes for the home. The widespread presence increased the chances that individuals would be drawn to natural history as a lifelong avocation. Not many devoted their energies beyond childhood, but those who did contributed in turn to the disciplines and the discourse.[4]

These Smithsonian correspondents, for example, demonstrate that Americans of varying classes and locales developed a profound love of natural history. They range in occupation from physicians, such as O. P. Baer of Richmond, Indiana, who "for pleasurable pastime ha[s] carried on the various Sciences—ha[s] been collecting ever since 1835," to men of independent wealth and (making up the bulk of respondents) farmers. The frequent rhetorical addresses to farmers in the agricultural press to gather up natural history objects had been heeded by men such as Silas West, "farmer & carpenter" in Cornish, Maine. He had "studied Geology, mineralogy, Astronomy, Botany, Meteorology and surveying a little so that I have a little Idea of them" and owned "about 100 kinds of Minerals and some birds and Beetles etc. I have not any good place to keep a cabinate [*sic*]." The wealthy retiree Robert L. Stuart, a subscriber to the Audubon and Elliot folio works who bankrolled the American Museum of Natural History in its early days, modestly listed his interests as illustrated travel and natural history books; the Maine businessmen George Boardman noted his huge collection of bird skins. The responses of a few correspondents suggest that women and those outside the traditional upper class were seriously interested: Miss Emma A. Limieth of Peoria, Illinois, listed her collection "of the Herpetology of Ill. and entomology," and Joseph Bell Franklin of Venango County, Pennsylvania, "quarryman," owned "a collection of Plants about 8 or 10 hundred specimens." (An itinerant minister in Nebraska admitted that his occupation made keeping collections impossible.) The geographical distribution, including Texas, Virginia, British Columbia, and even Alaska (one contributor), reflects not only the Smithsonian's desire for continentwide information but also widespread individual interest.[5]

The specific topics of interest demonstrate the diversity and breadth typical of the printed material of past decades. Many noted more than one scientific interest. B. F. Abell's interests, described by his wife in Troy, Ohio, as "Science, Philos., Animals, Birds, Insect life, Fruit, Flowers, Trees, Plants" with a cabinet of "Butterflies, moths . . . skulls of turtles, all the river shells, skulls in species of furred animals from Racoon to Bats, also of a few birds, 170 varieties of Birds Eggs"

222

The Book
of Nature
Natural History
in the
United States

were perhaps the most complete and wide-ranging. Many combined the disciplines connected with the earth: mineralogy, paleontology, and because of its association with fossils, conchology. Entomology and ornithology (with the subfield of oology, the study of eggs) were favorites, particularly for building collections. Most who pursued these disciplines expressed interest in the other biological sciences, such as mammalogy and botany. Ethnology is listed frequently, with collecting Indian relics the corresponding pastime, although only J. S. Patterson of Berlin, Ohio, connected his "especial interest in Anthropological studies" to his general interest for science to contribute to "human happiness." The most ambitious interest belonged to Seth Green, whose occupation was related to game-fish stocking: "to see all the waters in Whole World well Stock[ed] with fishes."[6]

The responses point to growing trends as well as past interests. A few, such as J. A. Allen of Harvard's Museum of Comparative Zoology, were connected to the young science museum movement. Two correspondents labeled themselves "amateurs," and many called their collections "of no interest," perhaps in deference to growing expertise of naturalist-authors like Allen, who made their living from studying their disciplines, and to the national collection-building efforts of the Smithsonian. A few correspondents such as John Batty and Martha Maxwell and lesser-known figures such as John B. Gilbert of Gates County in New York state, "Gun Dealer & Maker of Sporting Arms, Taxidermist," who had made a collection "of Mounted Birds 600 Native State different positions [sic]," supported themselves by hunting, selling, and stuffing birds and mammals. The presence of these "entrepreneurial naturalists" even in limited numbers signifies that they had customers building their own collections who were willing to purchase supplies and specimens. Demuth Brothers, a Brooklyn, New York, firm, supplied, presumably to such merchants, the artificial eyes for fish, birds, and mammals (it also manufactured dolls' eyes). The responses of several students, some studying to become science teachers, underline the expansion of the natural sciences in the classrooms of schools and colleges: Mary E. Andrews of Delaware, Ohio, was "preparing for Teacher of Natural Science," and W. Osburn, a student at Kansas State University in Lawrence, was spreading his knowledge through his editorship of the newspaper column "Observer of Nature." Others had given their cabinets to schools or colleges, among them O. P. Baer, who gave "$1/2$ ton of [fossil] specimens [to] Logansport College."[7]

The conjunction of occupations and interests underscores the expansion of individuals' intellectual and emotional horizons. The

various interests go well beyond practical considerations or financial
gain in most cases and so must have been adopted out of personal
satisfaction. As Baer noted, he and the others "for pleasurable pastime have carried on the various Sciences." A Wellesley, Massachusetts, real estate agent, Charles Blaisdell, who collected 1,000
specimens of fossils and shells, and the Calais, Maine, retiree George
Boardman, who "found 265 species in this locality," were seeking
new interests and challenges; their pride and ambition are evident
as they relate the size of their collections. The only regret that the
correspondents wistfully expressed was lack of time and resources to
pursue their love. Susan M. Wheaton near Independence, Iowa, who
listed herself as "House-Mother," stated, "Natural History is a subject
of much interest to me, but I have had very little leisure or chance to
pursue the study," so her "miscellaneous" collections of "mosses,
shells, [and] savage weapons" remained unclassified. Theodore Day,
owning a "rough place, 25 acres, $^1/_2$ cleared, no team, poor build-
ings," wished he was "able to travel some and collect, minerals, stone,
flowers &c Butterflies, beetles &c," but with limited means he could
never buy cabinets for his insect and mineral collections: "Work hard
for a living."[8]

No correspondent specifically credited particular literature in
forming his or her tastes. However, the emphases reflect wide-ranging
and eclectic interests fostered by textbooks and popular works of the
time. As the antebellum children's works, newspaper extracts, and
periodical articles spanned discussions of humanity, mammals, birds,
and insects as well as geology, so did these correspondents' interests.
Surely the "some 150 original drawings of birds mostly life
size — coloured" of farmer William Temple Allen of Clark County,
Virginia, were specifically inspired by the ornithological illustration
tradition. According to her husband, James Lewis—himself a noted
conchologist—Ida A. Lewis drew "finely and numerously" their collection of shells by showing their dorsal and ventral views as in scientific treatises. Peculiarly American biases are manifested in the
interest and collection of "savage artifacts" and relative lack of interest in microscopy as compared to the European practices. The correspondents studied their local fauna, fossils, and relics, for, as William
Anderson, a South Carolina physician, stated, "the Fauna and Flora
of the vicinity necessarily come most frequently under a passing observation." Thus, generalized interest in all natural production, fed
by reading and viewing, triggered specialization in collecting, such as
Limieth's "herpetology of Illinois" and Gilbert's stuffed "600 native
birds."

The correspondents' collecting in the absence of easy-to-use manuals for identification attests to their perseverance, especially to the modern observer perhaps spoiled by omnipresent nature guides. A few respondents pleaded with the Smithsonian to send catalogs to help them identify, classify, and organize collections of hundreds of insects, plants, and mammals, although the catalogs only listed the American fauna in Latinate terms with abbreviated phraseology. Botany continued to be popular in the classroom and in the field, as evidenced by the sales of new editions of Asa Gray's work, *Elements of Botany,* and his primer intended for very young beginners, *How Plants Grow,* well into the twentieth century. Botanists had textbooks such as Gray's and Lincoln's works, which served well as introductions to the discipline because they thoroughly addressed identification and terminology. In contrast, ornithology and entomology enthusiasts had to rely on the Audubon works or unillustrated Baird catalogs. Practitioners of entomology were forced to rely on scattered articles and scarce texts, such as Harris's *Injurious Insects,* until after the Civil War.

The publication of the first edition of Elliott Coues's *Key to North American Birds,* in 1872, presaged the twentieth-century outpouring of field guides, led by bird-watching titles. Illustrated with woodcuts derived from Audubon and by the author himself, Coues claimed to have organized the text to promote easy identification, as he boasted that his wife, who did not recognize birds, could work through the text to identify the red-headed woodpecker lying in front of her (note the dead specimen in the years before binocular use allowed precise sighting of live birds). The descriptions are much fuller and contain far more anatomical and historical information than today's familiar guides. Nonetheless Coues included enough species in a single, portable, well-illustrated volume to spur on "birders" in the work's five editions. Coues's success did inspire shorter and cheaper manuals, such as Chester A. Reed's pocket-size *Bird Guide,* which Roger Tory Peterson first studied as a youngster in the 1920s. Peterson himself would further popularize the genre in his field-guide series not only for birds but for wildflowers, which sold millions, by developing a system of identification dependent on recognizing solely visual characteristics, as opposed to the older discussions of families, genera, and species.[9]

Around the turn of the century, the Cornell professor John Henry Comstock, who had memorized difficult genera in Harris's *Injurious Insects* for his introduction to entomology, composed popular introductions to entomology with the aid of his wife. These were written in much simpler terms, outlining means for identification and providing detailed information for insect collecting. He and Anna Botsford Comstock geared their texts to teachers and students of the burgeoning "Nature Study" movement, whose advocates urged urban and rural youth to experience "Nature" and the outdoors directly. Comstock's *Insect Life* (first published in 1897) built chapters around habitats ("Pond Life," "Forest Life," and "Roadside Life"), as the new science of "ecology" was fostering interest in the interrelationships of species in a given environment. But, as with the rest of the introductory books and field guides, identification and subsequent knowledge of individual species were regarded as necessary to understanding the overarching "Nature" and interconnectedness of its communal life. Comstock begins the work with descriptions of genera and explanations of classification with special emphasis on life cycles, and in following chapters he instructs students to study individual species' habits and "write an account of your observations on these insects and illustrate it." Note that Comstock himself portrays his favorite pond in terms of its separate species and their characteristic movements: "There is a pond that we love to visit when we are tired at our desks. . . . It is a long, narrow one. . . . Over this pond dragon-flies hawk at midges; on a dead tree near the bank a kingfisher has his perch, from which on our approach he swoops down twirling his watchman's rattle; here, too, occur large colonies of whirligig beetles, which chase each other round and round as if at play."[10]

As was true for the earlier ornithologists, hours of observation informed by reading deepened not only empirical knowledge but also fostered a kinship beyond aesthetic appreciation: "What pleasure when one is tired to lie on a grassy bank and watch the ripples chase each other over the water. . . . Such experiences bring rest and a feeling of harmony with Nature. But a keener enjoyment comes with a more intimate acquaintance with the forms of life that abound in these places, when one can look upon each kind of water-plant as an old friend, and know something of the ways of the creatures that glide over the surface or swim underneath." The "friendship" of individual species makes the natural world of Comstock and his audience at "home" in the natural environment.[11]

The life history format outside the monograph would be reinvigorated in the post–Civil War years. Beginning in the 1870s a boom of periodicals devoted to nature observation such as *The Oologist* provided opportunities for authors to concentrate on species in articles like "The Mule Deer" and the "Land Snails of New England." Read-

226

THE BOOK
OF NATURE
Natural History
in the
United States

ers shared their own firsthand natural history observations with other subscribers in their letters to these magazines. The format structured Ernest Thompson Seton's *Life Histories of North American Animals* (1909) and Arthur C. Bent's monumental *Life Histories* of American bird families published in the *Bulletin of the United States Museum* throughout the first half of the century. Reports from the federal Biological Survey before its transformation to the Fish and Wildlife Services in 1940 incorporated the newer concerns of population dynamics and predator relationships into life histories.[12]

Increasing professionalism in the biological sciences in government agencies, research institutions, and universities, toward the end of the century encouraged attitudes that differed from those of the larger society. These newer representatives of natural history gradually expunged the older overt references to theology and appeals to readers' sentiments, as the scientific community reacted against the talking birds and insects of earlier children's literature and disputed humanlike intelligence as guiding behavior. Thomas Nuttall's praise of birds' "conjugal fidelity" and William Henry Edwards's admiration of insect reasoning would appear inappropriate to these authors. The influential "Morgan's Canon," expounded by C. Lloyd Morgan in 1894, for example, ascribes animal actions to the lowest denominator of psychological faculties rather than to higher intelligence.[13]

Around the turn of the century, popular nature writers flouted this "canon" and chose to intensify the closeness of animals to humanity in new fictional formats. Gene Porter Stratton's novels and Seton's short stories star animal protagonists; books such as Mabel Osgood Wright's popular *Citizen Bird* revitalize the familiar device of dialogue imparting the species' habits and family lives. The latter's cast of characters includes a naturalist-doctor and his young relatives, Olive, Nat, and Dodo, who enact the sentiment of keeping wild birds free by purchasing caged cardinals from a store and releasing them on "Liberation Day." As in the earlier works, knowledge of the full life history of a bird engenders appreciation.[14]

The most evident offshoot of emotional and graphic passages in Wilson's and Audubon's life histories is the short nature essay, the melding of personal factual observation and sentiment, usually published as general literature. Its most prolific American practitioner in the nineteenth and early twentieth century, John Burroughs, had pored over the Wilson and Audubon texts at West Point Academy as a young adult in his first exposure to a well-stocked natural history library. Thenceforth, he constantly regarded Audubon as an authority and often praised his "enthusiasm so genuine and purpose so single," despite lamenting the "verbose and affected style." He himself chose ornithology as his "avenue of delight" into nature, exclaiming in *Wake-Robin*, an anthology of his essays, "What a new delight the woods have!" before offering his more concrete advice: "[O]bserve the bird, then shoot it (not ogle it with a glass), and compare it with Audubon." Burroughs, unlike the older ornithologists, had to contend directly with the perception of cruelty and waste in "Nature," as suggested by Darwin's *Origins of Species*, yet the outdoors and the species under study remain the source of inspiration and beneficence for him.[15]

Lesser-known nature essayists such as Maurice Thompson, who published his meditations in the *Atlantic Monthly*, blatantly romanticized the Darwinian struggle for existence: "The skull of a bird is the flower of a long *chanson de geste* coming up through a million years of adventure and change from the fish or the saurian. Every species of plant or animal has an heroic significance when I remember what a battle it has fought for its existence." Clearly rejecting scientific materialism, he states a belief in evolution, "But it is evolution by God's laws, bounded by His limiting purpose." As recent anthologies have shown, the nature essay, often based on observation of species and reliant on vivid aural and visual description, has flourished up to the present in the writings of Edwin Way Teale, John Hay, and Ann Zwinger, to name but a few. Overt reference to God might now be omitted, but the belief in spirituality in "Nature" and the literature's dual appeal to intellect and emotions remain.[16]

Other Legacies

Technological developments in photomechanical reproduction provided opportunities for book illustration hitherto limited to the techniques of expensive hand-colored plates, monochrome wood engravings, and inexact chromolithography. For example, the Comstocks' works combined Anna Botsford's elegant wood engraving with illustrations using the four-color process from printing plates based on original photographs. Later natural history illustrators chose to follow the iconographic traditions, as the careers of several faunal artists demonstrate. The most widely published bird and mammal artist of the early twentieth century, Louis Agassiz Fuertes, developed a consistent imagery of the single animal "with a dash of the environment" and "suggestion of a haunt," as he described it, after intensely studying Audubon's work as a youth. His works were widely reproduced in state bird books, children's literature such as *Citizen Bird*,

228

THE BOOK
OF NATURE
Natural History
in the
United States

and advertising cards. Individuals such as Abbott Handerson Thayer, Francis Lee Jaques, and Sweden's Bruno Liljefors attempted to forge new imagery dynamically integrating flora and fauna into their habitat. Nonetheless, later nature artists who frequently illustrated books, such as Roger Tory Peterson and Don Eckelberry, continued to use the iconography of the species representative clearly delineated as the centerpiece, with background "suggested." These same artists, freed from illustration duties, favored this same approach in their original drawings and paintings, frequently reproduced in popular print editions. Other artists in the "nature" genre, such as Robert Bateman today, also present ideal specimens in their characteristic environs. While believing their greatest influence is "Nature," these wildlife artists work in the same conventions that filter and order their perceptions of their subjects as did their predecessors. Similarly, their audiences hold this conventionalized image of the wildlife creatures; viewers in the early part of the twentieth century, claimed Peterson, saw the bird "as Fuertes drew it, not as it actually looked in the brief moment it perched before [them]."[17]

Today's still and moving photography also portrays specimens against their characteristic environs (Illustration 6.1). For example, the "plant portrait" remains a staple of flora photography, as the ideal specimen is shot in its entirety against a background that is often in softer, less distinct focus. Manuals on wildlife photography note that the most popular subjects remain single animals or small groups of one species occupying the foreground, despite concern for interrelationships among species in a given habitat. They also emphasize the characteristic attributes of life histories: feeding, mating, birth, and aggression. The genre of wildlife films links imagery and narrative usually to the specimen standing for the species and frequently relies on the close-up with hazy background. The characteristic movements of catching prey, feeding young, and general locomotion only implied in still imagery now are seen in totality.[18]

Historians consistently have appreciated the role of nature writing in inspiring conservation legislature in the United States. More specifically but less emphasized, the dissemination of the natural history disciplines was the key to appreciation and protection of individual species and, beyond them, their surrounding landscape. The inclusion of natural history information in the daily press and major publishers' books signaled to literate Americans that flora and fauna, foreign and exotic, merited attention. The persistent presence of the life histories and the accompanying imagery, moreover, gave the species an autonomy apart from humans, for they had their own "histories." A familiarity with the surroundings, daily activities, and appearance different from that of everyday experience was engendered from earliest childhood. Such knowledge made possible the impetus toward conservation and preservation.

Natural history study of the two most widely disseminated disciplines in antebellum and post–Civil War America, botany and ornithology, triggered the nature appreciation of the movement's most influential members. The patron saint of the national parks movement, John Muir, began his career in the 1860s as an amateur botanist who also wrote bird essays such as "The Water-Ouzel," published in the *Atlantic Monthly*. John Burroughs, who in the early 1900s became an outspoken member of the bird preservation movement and

Illus. 6.1. Burrowing owl. Photograph by Herb Crossan, n.d. (Courtesy of Herb Crossan.)

230

THE BOOK
OF NATURE
Natural History
in the
United States

whose highly popular writings exuded respect for his neighboring woods and fields, similarly began his life's work by studying ornithology. The young Theodore Roosevelt, upon entering Harvard, had hoped to be a naturalist, "a scientific man of the Audubon, or Baird, or Coues type," but instead continued his bird-watching in the White House confines. The popular children's nature writers who strove to point out to their readers the virtues of protecting their subjects, Mabel Osgood Wright and Florence Merriam, were themselves ornithologists. George Bird Grinnell named his nascent wildlife-protection society after Audubon, "who did more to teach Americans about birds of their own land than any other who ever lived." As important as these major figures were the many individuals who read these authors, joined the early Audubon clubs, supported the various protective state and local laws, and understood and sympathized with the dual autonomy yet kinship with animal species, particularly birds, that the authors and lawmakers, in the tradition of earlier ornithologists, had sought to promote.[19]

The legislation and institutional actions themselves reflect the legacies of the life history tradition. Because of the preponderance of ornithologically inspired literature, the first faunal nongame species protected against human hunting were "harmless" and "insectivorous" songbirds, through several state laws in the 1840s and 1850s. The coalition of turn-of-the-century nature writers, Audubon groups, the American Ornithological Union, and humanitarian organizations, buoyed by the general public's sympathy for birds instilled since childhood, promoted a flood of state and federal legislation protecting against plume-hunting and egg and bird destruction. In the early 1900s the Roosevelt administration and the Audubon Society began to purchase and set aside habitat in continuing efforts to protect waterbird species. Future preservation societies would develop similar strategies as they realized that supporters empathized with species as if they were individuals—even friends—in peril. A series of laws protecting wildlife culminating in the Endangered Species Act of 1973 codified the preeminence of protecting individual species in their habitats, although recent environmental supporters might prefer a more straightforward protection of total ecosystems.[20]

The conjunction of individual interests and communal disciplines in natural history is unique to the time period 1825–1875, although it left powerful legacies. No longer would a commentary such as Mrs. P. A. Messer's account of feeding her pet Baltimore oriole from her mouth be solicited for a scientific text such as Thomas

Brewer's edition of *American Ornithology*. The scientific community's desire diminished for contributions such as those of the salamander eggs hunted by the Reverend Charles Mann's sons and the numerous reptiles dispatched by Mary Andrews to Baird at the Smithsonian. Not only did the appreciation of wildlife, fostered by natural history texts and didactic children's literature, slowly discourage the use of wild creatures as pets, but the scientific disciplines were establishing boundaries between amateurs and professionals. Entomology, botany, and ornithology continued to incorporate the findings of amateur collectors, but, in organizations and literature, leaders in the field mostly wrote and associated with other full-time academics or government-sponsored scientists. Animal and plant physiology and a host of other topics in the newly emerging science of biology overshadowed the older interests in classification and the instinct/intelligence duality to which nonprofessional observers were able to contribute anecdotes and samples. Scientific endeavors depended on governmental supports and educational institutions instead of the private supporters that the national monographs had relied on. The shared interest in natural objects no longer would unify contending elites, as was true of Alexander Wilson's *American Ornithology* patronage, or divert serious social and political concerns, as had been the case in antebellum America.[21]

Nor would an important government official such as Baird ever again write to a minister plenipotentiary and U.S. senator such as George Marsh, "By all means send me lots of Salamandrosus. I want him exceedingly." The central importance in literate culture of natural history practice and discourse, as manifested in the literature and lifestyle of Americans, had lessened. Natural history entered lives through childhood reading and education and often became the purview of female writers and educators, but other avenues to explore the outdoors were introduced as well. The interests and pastimes available to educated adults of both sexes had also multiplied since the earlier generation, lessening the likelihood of politician-naturalists like Marsh to encourage zoologists (the exceptions were the Roosevelt presidents, both bird-watchers). Men whose standing was comparable to that of the Audubon subscriber and natural history enthusiast Daniel Webster now held other opportunities for interesting and socially sanctioned activities beyond natural history, horticulture, and hunting. Many pursuits like photography and bicycling were open to women also, extending to their participants the pleasures of "Nature" and the outdoors, laden with quasi-religious values.[22]

232

THE BOOK
OF NATURE
Natural History
in the
United States

As other students of Victorian culture have commented, the attitudes arising from the antebellum and post–Civil War ferment and expansion vividly and unexpectedly persist in today's culture despite technological change and temporal distance.[23] The absorption in a natural place throughout its yearly cycle, the respect for a species' own life cycle and unique adaptations, and the interest in bird, animal, and even plant behavior akin to human actions—themes expressed in natural history and potently disseminated through literature and imagery between 1825 and 1875—emerge today in forms as diverse as published journals reverently observing the changing seasons and television news snippets extolling the rescue of sea mammals. Expressions of concern for diminution of nonhuman life forms in the nineteenth-century conventional literature could be applied verbatim to today's issues. Although it is tempting to distinguish ourselves as more enlightened and therefore potentially less destructive than the nineteenth-century leaders whose deeds transformed wildlife and habitats, the early and frequent lamentations assure us that many were fully conscious of the toll their actions took. Awareness alone did not preserve the newly cherished flora and fauna.

Notes

INTRODUCTION

1. "Audubon's Ornithology, First Volume," *American Journal of Sciences and Arts,* 1st ser., 39 (1840): 348.

2. As quoted in Anna Botsford Comstock, *The Comstocks of Cornell: John Henry Comstock and Anna Botsford Comstock,* ed. Glenn W. Herrick and Ruby Green Smith (Ithaca, N.Y.: Comstock Publishing Associates, 1953), 34.

3. W. L. Strain to John James Audubon, 25 November 1845, John James Audubon Papers, Gen. MSS. 85, Beinecke Rare Book and Manuscript Library, Yale University (hereafter cited as Audubon Papers, Beinecke Library).

4. Daniel Chester French, introduction to William Brewster, *October Farm: From the Concord Journals and Diaries of William Brewster* (Cambridge, Mass.: Harvard University Press, 1936), ix–x; Charles Foster Batchelder, *An Account of the Nuttall Ornithological Club, 1873 to 1919,* Memoirs of the Nuttall Ornithological Club, no. 8 (Cambridge, Mass.: The Club, 1937), 11.

5. A. B. Comstock, *Comstocks of Cornell,* 33–34.

6. Peter A. Fritzell, *Nature Writing and America: Essays upon a Cultural Type* (Ames: Iowa State University Press, 1990), 137, in an interesting discussion of how seventeenth-century naturalist John Bannister dispassionately described an insect invasion illustrates how the settlers employed the natural history discourse to explain the disturbing new phenomena they encountered. Furthermore, he suggests that the exploration of objects and phenomena provided a necessary psychic relief. (See also chap. 1, n. 10.) Gary Paul Nabhan, in "Field Notes and the Literary Process," in Edward Lueders, *Writing Natural History: Dialogues with Authors* (Salt Lake City: University of Utah Press, 1989), 71, echoes this thought in a more contemporary context: "I think it [practicing natural history] helps each of us as individuals gain perspective on the 'external' world. A focus on something other than ourselves is probably healthy in this society, because we are so consumed with ourselves."

7. I loosely use the term *intensive* to signify repeated readings with a potential impact on the reader's outlook and actions and not in the classic sense of *intensive* versus *extensive,* forwarded by Rolf Engelsing. Robert Darnton's portrayal of Voltaire's readers incorporating his novels into their daily lives in "Readers Respond to Rousseau," in *The Great Cat Massacre and Other Episodes in French Cultural History* (New York: Basic Books, 1984; rpt. New York: Vintage Books, 1985), 215–56, exemplifies in my view "intensive reading."

8. I have relied most upon the discussions and definitions of literate antebellum and post–Civil War culture in Anne C. Rose, *Victorian America and the Civil War* (New York: Cambridge University Press, 1992); Karen Halttunen, *Confidence Men and Painted Women: A Study of Middle-Class Culture in America, 1830–1870* (New Haven, Conn.: Yale University Press, 1982); Nancy Dunlap Bercaw, "Solid

Objects/Mutable Meanings: Fancywork and the Construction of Bourgeois Culture, 1840–1880," *Winterthur Portfolio* 26 (1991): 232–47; Ann Douglas, *The Feminization of American Culture* (New York: Alfred A. Knopf, 1977); Daniel Howe, "Victorian Culture in America," in *Victorian America,* ed. Daniel Howe (Philadelphia: University of Pennsylvania Press, 1976), 3–28; and Lawrence W. Levine, *Highbrow/Lowbrow: The Emergence of Cultural Hierarchy in America* (Cambridge, Mass.: Harvard University Press, 1988). Bercaw's discussion of Raymond Williams's concept of a "common culture" in creating the nineteenth-century bourgeois ideology is particularly helpful (p. 239); two works specializing in the diffusion of print culture in the nineteenth century are Richard D. Brown, *Knowledge Is Power: The Diffusion of Information in Early America 1700–1865* (New York: Oxford University Press, 1989), and Ronald J. Zboray, *A Fictive People: Antebellum Economic Development and the American Reading Public* (New York: Oxford University Press, 1993). Ronald J. Zboray and Mary Saracino Zboray, "Books, Reading, and the World of Goods in Antebellum New England," *American Quarterly* 48 (1996): 587–622, analyze how a historical readership (in this case, families located in New England) bestows meanings on mass-produced materials.

9. William Martin Smallwood and Mabel Smallwood, *Natural History and the American Mind* (New York: Columbia University Press, 1941), 249–336, discuss the entrance of the biological disciplines into common schools, college curricula, lyceum lectures, and institutions. Max Meisel, *A Bibliography of American Natural History: The Pioneer Century, 1769–1865* (Brooklyn, N.Y.: Premier Publishing, 1924), vols. 1 and 2, gives short histories of the major and minor institutions. One discussion of the problems that early scientific organizations faced is Simon Baatz, "Philadelphia Patronage: The Institutional Structure of Natural History in the New Republic," *Journal of the Early Republic* 8 (1988): 111–38. Patsy A. Gerstner, "The Academy of Natural Sciences of Philadelphia, 1812–1850," in *The Pursuit of Knowledge in the Early American Republic: American Scientific and Learned Societies from Colonial Times to the Civil War,* ed. Alexandra Oleson and Sanborn C. Brown (Baltimore: Johns Hopkins University Press, 1976), 174–93, traces a difficult time in an important natural history institution's history. American historians of science have tended to neglect the popularization of natural history in their concentration on professionalization, but the publication of Elizabeth B. Keeney, *The Botanizers: Amateur Scientists in Nineteenth-Century America* (Chapel Hill: University of North Carolina Press, 1992), signals a change in direction. Earlier, Donald Zochert in "Science and the Common Man in Antebellum America," *Isis* 65 (1974): 448–73, explored the dissemination of the disciplines in midwestern newspapers. David Elliston Allen, *The Naturalist in Britain: A Social History* (London: Allen Lane, 1976), describes institutions such as the natural history clubs that formed in considerable numbers in the United States after 1875 (see Keeney, *Botanizers,* 79, 140–41.)

10. The survey works on the European tradition of natural history, particularly on illustration, are legion. Some of the more useful are David Knight, *Natural Science Books in English, 1600–1900* (New York: Praeger, 1972; London: Portman Books, 1989); S. Peter Dance, *The Art of Natural History: Animal Illustrators and Their Work* (Woodstock, N.Y.: Overlook Press, 1978); Joseph Kastner and Miriam T. Gross, *The Animal Illustrated, 1550–1900: From the Collections of the New York Public Library* (New York: Harry N. Abrams, 1991); Madeleine Pinault, *The Painter as Naturalist from Dürer to Redouté,* trans. Philip Sturgess (Paris: Flammarion, 1991);

and A. M. Lysaght, *The Book of Birds: Five Centuries of Bird Illustration* (1975; rpt., New York: Exeter Books, 1984). Botanical illustration is represented in Ann Arber, *Herbals: Their Origin and Evolution*, 2nd ed. (London: Cambridge University Press, 1938); Wilfrid Blunt and Sandra Raphael, *The Illustrated Herbal* (London: Thames & Hudson, 1979); Bernard McTigue, *Nature Illustrated: Flowers, Plants, and Trees: Illustrations 1550–1900: from the Collections of the New York Public Library* (New York: Harry N. Abrams, 1989); and John V. Brindle and James J. White, *Flora Portrayed: Classics of Botanical Art from the Hunt Institute Collection* (Pittsburgh: Hunt Institute for Botanical Documentation, 1985).

11. The definitive work on U.S. zoological illustration is Ann Shelby Blum, *Picturing Nature: American Nineteenth-Century Zoological Illustration* (Princeton, N.J.: Princeton University Press, 1993). Blum comments on the relationship of image to text in her introduction, pp. 6–8, and discusses the iconography of the "living animal" as opposed to the more diagrammatic and anatomical format throughout the text beginning with p. 45. She points out that illustration is often subordinate to the text in that the text introduces the illustration to the reader and often exists without specific illustration. She and other historians of science who study imagery, such as Martin J. S. Rudwick in *Scenes from Deep Time: Early Pictorial Representations of the Prehistoric World* (Chicago: University of Chicago Press, 1992), imply that both text and imagery are usually complementary in that they discuss the same theory and principles. However, David Knight in the chapter "Discourse in Pictures," in his *The Age of Science: The Scientific World-View in the Nineteenth Century* (New York: Basil Blackwell, 1986), 109–24, notes that illustrations often are used for more than one text for several decades, thereby lessening the direct connection between text and illustration. There is no similarly inclusive work devoted to American botany works. John C. Greene, *American Science in the Age of Jefferson* (Ames: Iowa State University Press, 1984), and George H. Daniels, *American Science in the Age of Jackson* (New York: Columbia University Press, 1968), provide a general background up to 1850.

12. Albert Boime, *The Magisterial Gaze: Manifest Destiny and American Landscape Painting c. 1830–1865* (Washington, D.C.: Smithsonian Institution Press, 1991), and Angela Miller, *The Empire of the Eye: Landscape Representation and American Cultural Politics, 1825–1875* (Ithaca, N.Y.: Cornell University Press, 1993), are two examples that define a focus, in this case the panoramic view in landscape painting, in order to study a period's attitude. Echoing the nineteenth-century commentators, Blum, *Picturing Nature*, 41–42, likens Audubon's illustration to Charles Wilson Peale's portraits of Revolutionary War heroes. Keith Thomas in *Man and the Natural World: A History of the Modern Sensibility* (New York: Pantheon, 1983), 66–67, suggests the implicit autonomy given nonhuman species in natural history discourse.

13. Miller, *Empire of the Eye*, 20, 167–69, convincingly discusses the importance of localism in contrast to nationalistic programs in landscape painting.

14. Ibid., particularly pp. 209–41.

15. Two compilations that may serve as an introduction to U.S. historical readership are James L. Machor, ed., *Readers in History: Nineteenth-Century American Literature and the Contexts of Response* (Baltimore: Johns Hopkins University Press, 1993), and Cathy N. Davidson, ed., *Reading in America: Literature and Social History* (Baltimore: Johns Hopkins University Press, 1989). In comparison, there is less literature on historical "reader/viewer" response, but Joachim Möller, ed., *Imagi-*

nation on a Long Rein: English Literature Illustrated (Marburg: Jonas Verlag, 1988), contains essays relating to interpretation of text and illustration. Robert N. Essick, "Visual/Verbal Relationships in Book Illustration," in *British Art 1740–1820: Essays in Honor of Robert R. Wark*, ed. Guilland Sutherland (San Marino, Calif.: Huntingdon Library, 1992), 169–204, discusses the difficulties in using the same terminology to analyze both illustration and text, for discursive theory, developed from language criticism, does not relate easily to visual imagery in his opinion ("most people have an intuitive and deeply held sense that there are fundamental differences between pictures and words," 186).

16. Zochert, "Science and the Common Man," notes the intense interest in astronomical events, 450–51.

17. Perry Miller, "Nature and the National Ego," in *Errand into the Wilderness* (Cambridge: Belknap Press of Harvard University Press, 1956), 204–16, and *Nature's Nation* (Cambridge: Belknap Press of Harvard University Press, 1967); E. Douglas Branch, *The Sentimental Years, 1836–1860* (New York: D. Appleton-Century, 1934), 146.

18. Studies such as Vera Norwood, *Made from This Earth: American Women and Nature* (Chapel Hill: University of North Carolina Press, 1993), are creating awareness of women's interest in nature as evinced in illustration, writing, and gardening. Ann B. Shteir, *Cultivating Women, Cultivating Science: Flora's Daughters and Botany in England, 1760–1860* (Baltimore: Johns Hopkins University Press, 1996), makes the strong argument that women were guided to the margins of scientific culture and permitted limited participation through illustration and educational writing. The comparative weakness of the U.S. institutions and scientific community suggests that this argument applies to a lesser degree to this same time period and is better applied to later decades. In other words, U.S. naturalists and their societies were not a force united in their exclusion of women and perhaps needed female participation more than their British counterparts. Moreover, it should be remembered that very few Americans, male or female, during this period demonstrated the long-term desire to become naturalists.

CHAPTER 1

1. Alexander Wilson, "Proposals for Publishing by Subscription, . . . *American Ornithology*," [1807] (hereafter cited as "Prospectus") in *The Life and Letters of Alexander Wilson*, ed. Clark Hunter, Memoirs of the American Philosophical Society, vol. 154 (Philadelphia: American Philosophical Society, 1983), 270.

2. The standard works explicating the difficulties faced by artists, authors, and their supporters in building patronage of the arts include Lillian B. Miller, *Patrons and Patriotism: The Encouragement of the Fine Arts in the United States, 1790–1860* (Chicago: University of Chicago Press, 1966); Neil Harris, *The Artist in American Society: The Formative Years, 1790–1860*, rev. ed. (Chicago: University of Chicago Press, 1982); and William Charvat, *Literary Publishing in America, 1790–1850* (Philadelphia: University of Pennsylvania Press, 1959). William Charvat, in *The Profession of Authorship in America, 1800–1870: The Papers of William Charvat*, ed. Matthew Bruccoli (Columbus: Ohio State University Press, 1968), 9, relates that Joel Barlow found 769 subscribers to the 1787 edition of *The Vision of Columbus* (it should be noted that copies were sold for under two dollars).

3. Wilma George, *Animals and Maps* (Berkeley: University of California Press, 1969); George Brown Goode, "The Beginnings of Natural History in

America," in *The Origins of Natural Science in America: The Essays of George Brown Goode*, ed. Sally Gregory Kohlstedt (Washington: Smithsonian Institution Press, 1991), 44–45. Goode includes lesser-known French, Spanish, and Dutch exploration texts; Elsa Guerdum Allen in her *History of American Ornithology before Audubon*, Transactions of the American Philosophical Society, n.s. 41, pt. 3 (Philadelphia: American Philosophical Society, 1951), gives an extensive background on colonial travelers' lists that extend beyond birdlife. While I have chosen to lump together accounts of travelers from a variety of periods and backgrounds because of their strong stylistic consistency, Wayne Franklin in *Discoverers, Explorers, Settlers: The Diligent Writers of Early America* (Chicago: University of Chicago Press, 1979) elaborates his typology dividing the travel literature into discovery, exploratory, and settlement narratives.

4. One insightful article revealing the frequent adaptation of encyclopedia illustrations is William B. Ashworth Jr., "The Persistent Beast: Recurring Images in Early Zoological Illustration," in *The Natural Sciences and the Arts: Aspects of Interaction from the Renaissance to the 20th Century*, ed. Allan Ellenius (Uppsala: Almquist & Wiksell International, 1985), 46–66. For the history of herbals, see Arber, *Herbals: Their Origin and Evolution;* [John Gerard], *The Herball or General Historie of Plantes. Gathered by John Gerard. . . . Very Much Enlarged and Amended by Thomas Johnson* (London, 1633), 1516. Johnson explains in his own addition to Gerard's text about the plaintain that a friend had given him the plant "on April 10, 1633," thus emphasizing the accuracy of the image.

5. I have found particularly helpful Pamela Regis's definition of natural history discourse stressing listing and atemporality in *Describing Early America: Bartram, Jefferson, Crèvecoeur, and the Rhetoric of Natural History* (De Kalb: Northern Illinois University Press, 1992), 12–22, although she discusses mostly the "age of Linnaeus." Another author who analyzes the implications of natural history illustration's viewpoint and stylistics is Alex Potts, "Natural Order and the Call of the Wild: The Politics of Animal Picturing," *Oxford Art Journal* 13 (1990): 12–33; Wolfgang Harms, "On Natural History and Emblematics in the 16th Century," in Ellenius, ed., *Natural Sciences and the Arts*, 71–72.

6. Regis, *Describing Early America*, 23–25; she, in turns, cites literary criticism of ethnological literature such as Mary Louise Pratt, "Scratches on the Face of the Country," *Critical Inquiry* 12 (Autumn 1985): 120–21, as informing her discussion of timelessness and uniformity. Pratt's more recent work, *Imperial Eyes: Travel Writing and Transculturation* (New York: Routledge, 1992), is also pertinent.

7. For example, Kevin R. McNamara relates the changes in ornithological discourse to internal scientific development and the varied individual personalities in "The Feathered Scribe: The Discourses of American Ornithology before 1800," *William and Mary Quarterly* 47 (1990): 210–34.

8. I am indebted to Regis, *Describing Early America*, for her connection between ethnographic criticism and natural history (see n. 5 above). Franklin in his *Discoverers, Explorers, and Settlers* focuses on the qualities of travel literature, but speaks about the "struggle to express" the new plants and animals, 4–5. Among the many other discussions of the "other" related to colonial exploration is Tzvetan Todorov, *The Conquest of America: The Question of the Other*, trans. Richard Howard (New York: Harper & Row, 1984).

9. James R. Masterson, "Travelers' Tales of Colonial Natural History," *Journal of American Folklore* 59 (1946): 51–57, 174–88; see Percy Adams, *Travelers and*

Travel Liars, 1660–1800 (Berkeley: University of California Press, 1962), for an overview of the literature.

10. Regis, *Describing Early America,* expands on the "other" in reference to natural history, 23, 76. Fritzell, in *Nature Writing and America,* also discusses the need of "every zoologist or botanist" and "almost every landed American settler . . . to make his nonhuman others as distinctly other as possible," 137. For uneasiness shown by "odd" animals, see Angus K. Gillespie, "The Armadillo," in *American Wildlife in Symbols and Story,* ed. Jay Mechling and Angus K. Gillespie (Knoxville: University of Tennessee Press, 1987), 101–2. The Burke quote is from Regis, *Describing Early America,* 64–65.

11. William Bartram has probably attracted more attraction than any other colonial or early republic naturalist except Charles Willson Peale. A welcome new addition is Thomas P. Slaughter, *The Natures of John and William Bartram* (New York: Alfred A. Knopf, 1996). Scholarship in recent decades includes Amy R. Weinstein Meyers, "Sketches from the Wilderness: Changing Conceptions of Nature in American Natural History Illustration, 1680–1880," Ph.D. diss., Yale University, 1985, 113–93; Christopher Looby, "The Constitution of Nature: Taxonomy as Politics in Jefferson, Peale, and Bartram," *Early American Literature* 22 (1987): 252–73; Franklin, *Discoverers,* 59–76; Kerry S. Walters, "The 'Peaceable Disposition' of Animals: William Bartram on the Moral Sensibility of Brute Creation," *Pennsylvania History* 56 (July 1989): 157–76; Fritzell, *Nature Writing and America,* 138–44; Bruce Silver, "William Bartram's and Other Eighteenth Century Accounts of Nature," *Journal of the History of Ideas* 39 (1978): 597–614; and Regis, *Describing Early America,* 41–78.

12. William Bartram, *Travels through North and South Carolina, Georgia, East and West Florida* (London, 1792; facsimile edition, Charlottesville: University of Virginia Press by arrangement with the Beehive Press, 1980), 272. The work was first published in 1791 in Philadelphia.

13. For Lawson's life and work, I refer to John Lawson, *A New Voyage to Carolina,* ed. Hugh Talmage Lefler (Chapel Hill: University of North Carolina Press, 1967).

14. Mark Catesby, *The Natural History of Carolina, Florida, and the Bahama Islands* (London, 1731–43), I:v. A useful facsimile edition is *The Natural History of Carolina, Florida, and the Bahama Islands,* introduction by George Frick and notes by Joseph Ewan (Savannah: Beehive Press, 1974). The most complete biography on Catesby remains George Frick and Raymond Phineas Stearns, *Mark Catesby: The Colonial Audubon* (Urbana: University of Illinois Press, 1961). The work of Amy R. Weinstein Meyers, including "Sketches from the Wilderness," 46–112, contains perhaps the most extensive discussion of Catesby's distinctive artistic style. Her introductory essay in Henrietta McBurney, *Mark Catesby's Natural History of America: The Watercolors from the Royal Library Windsor Castle* (London: Merrell Holberton, 1997), 11–28, is also pertinent.

15. Bartram, *Travels,* 297, 280–94.

16. Ibid., 316, 122–23.

17. John Lawson, Letter to James Petiver, "Xber. 30th 1710," *New Voyage,* 270–71.

18. For biographical information on Abbot, see Vivian Rogers-Price, *John Abbot in Georgia: The Vision of a Naturalist Artist (1751–ca. 1840)* (Madison, Ga.: Madison-Morgan Cultural Center, 1983), and Marcus B. Simpson Jr., "Artistic

Sources for John Abbot's Watercolor Drawings of American Birds," *Archives of Natural History* 20 (1993): 197–212; Rogers-Price, *John Abbot in Georgia*, 92.

19. Rogers-Price, *John Abbot in Georgia*, 106, 39.

20. Abbot's modesty is evident in this letter to British naturalist William Swainson, 7 June 1819, Swainson Correspondence, Linnaean Society, London: "I am no botanist, but only an admirer of Nature's beauties; to meet with a new growing flower or plant much pleases me."

21. "Prospectus," *Life and Letters*, 270. The most complete volume available on Wilson is Clark Hunter, ed., *The Life and Letters of Alexander Wilson*, in which the entire prospectus dated 6 April 1807 is given, pp. 267–72. Robert Cantwell in his *Alexander Wilson: Pioneer and Ornithologist* (Philadelphia: Lippincott, 1961) eloquently describes Wilson's achievement, 253–54.

22. The quotation is from Alexander Wilson, "The Solitary Tutor," in *Poetical Works of Alexander Wilson, the American Ornithologist* (Belfast: John Henderson, 1844), 185.

23. See Introduction, n. 6.

24. Wilson, Letters to William Bartram, 10 November and 31 March 1803, *Life and Letters*, 204, 210–11; another letter, 17 November 1803, 204–5, refers to the copying of engravings: "I have taken the liberty of sending you another specimen of an attempt to imitate your beautiful Engravings."

25. Alexander Wilson, *American Ornithology* (Philadelphia: Bradford & Inskeep, 1808–14), 1: 56, 1:30; Wilson to William Bartram, 8 April 1807, *Life and Letters*, 261, on the unsuccessful nuthatch hunt; Letter to William Duncan, 24 December 1804: "I was frequently obliged to keep before [his two companions] and sing some lively ditty; to drown the sounds of the 'ohs!' and 'ahs!' and 'O Lords!'" *Life and Letters*, 227; *American Ornithology*, 3:80.

26. Wilson, *American Ornithology*, 1:56–57.

27. Ibid., 2:162.

28. Ibid., 2:87, 2:30. The "woman in the house" may have been a member of the William Jones family, with whom Wilson was living in 1803; "Prospectus," 269.

29. Wilson, *American Ornithology*, 2:149 and 4:62–63.

30. Ibid., 2:153; 4:97, 95; 2:41.

31. Wilson to William Bartram, 29 April 1807, *Life and Letters*, 262, in which Wilson discusses the adjective *migratorius;* Wilson, "Prospectus," 271.

32. Wilson, *American Ornithology*, 1:61; 1:50; 2:39–40; 1:104.

33. Wilson to Mordecai Churchman, 4 November 1811, *Life and Letters*, 394–95.

34. Wilson, *American Ornithology*, 5:53–56; 3:104–12; 2:154–58; [Nathan Winslow], *Dictionary of American Biography*, s.v., "Nathaniel Potter." Potter first corresponded with Wilson by answering his plea for information on the cow bunting published in the *Portfolio*. See "Correspondence," *Portfolio*, 3rd ser. 2 (1809), 151–53.

35. Wilson, *American Ornithology*, 2:100.

36. Ibid., 3:80, 1:107; Blum, *Picturing Nature*, gives the most complete analysis of Wilson's artistic style and its relationship to eventual published engraving and also discusses Charlotte Porter's acknowledgment of the introduction of landscape into natural history illustration, 33–46, 92. She acknowledges that Wilson did not consistently insert backgrounds throughout *Ornithology*.

37. Blum, *Picturing Nature*, 39.

38. Alexander Wilson to Alexander Lawson, 12 March 1804, *Life and Letters*, 207; Cathy N. Davidson, "The Life and Times of *Charlotte Temple*," in *Reading in America*, 165, on this novel's price in 1794; two advertisements by itinerant painters in 1806 Charleston (a city that Wilson visited) listed prices for oil paintings and miniatures between twenty and one hundred dollars. See Anna Wells Rutledge, *Artists in the Life of Charleston*, Transactions of the American Philosophical Society, n.s. 39, pt. 2 (Philadelphia: American Philosophical Society, 1949), 129; Judy L. Larson, "Dobson's Encyclopaedia: A Precedent in American Engraving," in Gerald W. R. Ward, ed., *The American Illustrated Book in the Nineteenth Century* (Winterthur, Del., and Charlottesville, Va.: Henry Francis du Pont Winterthur Museum distributed by the University Press of Virginia, 1987), 21–51. Larson notes that Dobson's work finally cost over $156 (p. 24), and the American edition of *Rees Cyclopedia* sold between three and five dollars a part (p. 45).

39. The buffalo in the frontispiece for the *Massachusetts Magazine* 4 (April 1792) may have been derived from "The American Bison" in Thomas Pennant's *History of Quadrupeds*, 3rd ed. (1793), reproduced in Dance, *Art of Natural History*, 83. Pennant's image, however, lacks the widespread eyes and human nostrils and mouth.

40. Wilson, Journal Extracts [1808], *Life and Letters*, 292; Letter to Daniel Miller, 5 March 1809, ibid., 305.

41. See Cantwell, *Wilson*, 277–305, for the list of subscribers with brief biographies.

42. Wilson, *American Ornithology*, 3:v; Jean V. Matthews, *Toward a New Society: American Thought and Culture 1800–1830* (Boston: Twayne Publishers, 1990), 5–6, gives a brief introduction on republicanism.

43. Wilson, *American Ornithology*, 1:40; Wilson, Letter to William Dunbar, 24 June 1810, *Life and Letters*, 375.

44. Wilson, *American Ornithology*, 2:viii.

45. Ibid., 5:75; 3:60, 5:35–36.

46. One edition of the popular work is [James Huddleston Wynne], *Choice Emblems, Natural, Historical, Fabulous, Moral, and Divine, for the Improvement and Pastime of Youth* (Philadelphia: Printed and Sold by Joseph Crukshank, 1790).

47. Ibid., 217, 48; for background on eighteenth-century humanitarism, see Dix Harwood, *Love for Animals and How It Developed in Great Britain* (New York, 1928), and James Turner, *Reckoning with the Beast: Animals, Pain, and Humanity in the Victorian Mind* (Baltimore: Johns Hopkins University Press, 1980), 1–14.

48. Wilson, *American Ornithology*, 4:98.

49. Bartram, *Travels*, 227–28; Wilson, *American Ornithology*, 1:15.

50. Samuel Williams, *Natural and Civil History of Vermont*, 2nd ed. (Burlington, Vt.: Printed by Samuel Mills, 1809), 1:479–82.

51. Jeremy Belknap, *History of New-Hampshire* (Boston: Bradford & Read, 1818), 3:73–96, 136. The origin of the cockroach had been a matter of concern for a number of naturalists. See Joseph Kastner, *A Species of Eternity* (New York: Alfred A. Knopf, 1977; New York: E. P. Dutton, 1978), 205; Williams, *History of Vermont*, 1:140–41.

52. I find helpful Kenneth John Myers's discussion of "disinterested knowledge" in a world of competing interests in his "On the Cultural Construction of Landscape Experience: Contact to 1830," in *American Iconology: New Approaches to Nineteenth-Century Art and Literature*, ed. David C. Miller (New Haven: Yale Univer-

sity Press, 1993), 64–65; McNamara, "Feathered Scribe," 212, calls many similar colonial accounts "discourses of commodification."

53. John Drayton, *A View of South-Carolina as Respects Her Natural and Civil Concerns* (Charleston, S.C.: W. P. Young, 1802), 38, 60–84, 88; Belknap, *History of New-Hampshire*, 125.

CHAPTER 2

1. Perry Miller, *The Life of the Mind in America from the Revolution to the Civil War* (New York: Harcourt, Brace & World, 1965), Harlan quoted on p. 286.

2. Charlotte Porter, in her important work, *The Eagle's Nest: Natural History and American Ideas, 1812–1842* ([Tuscaloosa]: University of Alabama Press, 1986), 57, notes, "Within a period of twenty-five years, these authors created an illustrated library of American natural history." Blum, *Picturing Nature*, pp. 47–48, cites *American Ornithology* as an exemplar. Historians have long acknowledged the American focus on discovery and cataloging of native species. See, for example, Greene, *American Science in the Age of Jefferson*, and Daniels, *American Science in the Age of Jackson*.

3. Constantine S. Rafinesque, *Medical Flora* (Philadelphia: Atkinson & Alexander, 1828), 1:vii, quoted in Georgia B. Barnhill, "The Publication of Illustrated Natural Histories in Philadelphia, 1800–1850," in Ward, ed. *American Illustrated Book in the Nineteenth Century*, 70.

4. Paul Lawrence Farber, *The Emergence of Ornithology as a Scientific Discipline, 1760–1850* (Dordrecht, Neth.: D. Reidel, 1982); Mark Velpeau Barrow Jr., "Birds and Boundaries: Community, Practice, and Conservation in North American Ornithology, 1865–1935," Ph.D. diss., Harvard University, 1992, 2–3, 15.

5. See Andrea J. Tucher, *Natural History in America, 1609–1860: Printed Works in the Collections of the American Philosophical Society, the Historical Society of Philadelphia, the Library Company of Philadelphia* (New York: Garland, 1985), 205–6, for a précis of the Wilson editions; Charles Lucien Bonaparte, *American Ornithology; Or the Natural History of Birds Inhabiting the United States Not Given by Wilson* (Philadelphia: Carey, Lea, & Carey, 1828), 2:23.

6. "Wilson N. A. Ornithology 1832," manuscript, Sir William Jardine, vol. 3, Jardine Hall Mss., Archives, British Museum of Natural History; Thomas Brewer, ed., *Wilson's American Ornithology: With Additions Including the Birds Described by Audubon, Bonaparte, Nuttall and Richardson* (Boston: Otis, Broaders & Co., 1840; reprint, New York: Arno Press, 1970).

7. The literature on Audubon is abundant. Some of the more complete bibliographies are in Francis Hobart Herrick, *Audubon the Naturalist*, 2nd ed. (New York: D. Appleton-Century, 1938; rpt. New York: Dover, 1968), 2:401–61, Alice Ford, *John James Audubon: A Biography* (New York: Abbeville Press, 1988), 501–10; Ron Tyler, *Audubon's Great National Work: The Royal Octavo Edition of* The Birds of America (Austin, Tex.: W. Thomas Taylor, 1993), 191–204. See John James Audubon, *My Style of Drawing Birds*, eds. Michael Zinman (Ardsley, N.Y.: Overland Press for the Hadyn Foundation, 1979), and Reba Fishman Snyder, "Complexity in Creation: A Detailed Look at the Watercolors for *The Birds of America*," in *John James Audubon: The Watercolors for* The Birds of America, ed. Annette Blaugrund and Theodore E. Stebbins Jr. (New York: Villard Books and New-York Historical Society, 1993), 55–68, for discussions of his unique artistic methods.

8. Robert Mengel, review of *The Original Water-Color Paintings of John James Audubon for* The Birds of America, ed. Marshall Davidson, *Scientific American* 216

(1967): 155–58; Blum, *Picturing Nature,* 112–13; and Amy R. W. Meyers, "Observations of an American Woodsman: John James Audubon as Field Naturalist," in *John James Audubon,* ed. Blaugrund and Stebbins, 49–51, analyze the narrative content in Audubon's artistic style. Carole Slatnick, in her catalog entries for *John James Audubon,* sensitively points out how Audubon meshes the graphic design with the species' characteristic markings and habits.

9. John James Audubon, *The Birds of America* (Philadelphia, 1840–1844; rpt. New York: Dover, 1967), 6:100–101. This octavo edition includes the original *Ornithological Biography* text. Audubon reports, "My wife admired them much on account of their gentle deportment, for although being tormented, they would spread their wings, ruffle their feathers, and draw back their head as if to strike, yet they suffered themselves to be touched by any one without pecking at his hand"; (London) *Athenaeum,* 10 May 1834, 350.

10. Robert Henry Welker, *Birds and Men: American Birds in Science, Art, Literature, and Conservation, 1800–1900* (Cambridge, Mass.: MIT Press, 1955), 86.

11. Robert Ralph, *William MacGillivray* (London: The Natural History Museum and HMSO, 1993), 82–83. Ralph also gives fascinating insights into the Audubon-MacGillivray relationship and reproduces MacGillivray's own exquisite bird drawings; *Birds of America,* 5:281.

12. Quoted in Ford, *John James Audubon,* 498; "Ornithological Biography," *American Quarterly Review* 20 (December 1831): 249.

13. Audubon, *Birds of America,* 4:191.

14. Thomas Nuttall, *A Manual of the Ornithology of the United States and of Canada,* 2 vols. (Cambridge: Hilliard & Brown, 1832, 1834). The quotations come from the revised edition of 1840 published by Hilliard, Gray, & Company of *The Land Birds* (vol. 1) written after his Rocky Mountain expedition, 206, 433, 715. (The second volume, dedicated to water birds, was never revised). The most complete biography is Jeannette E. Graustein, *Thomas Nuttall, Naturalist: Explorations in America, 1808–1841* (Cambridge: Harvard University Press, 1967).

15. Porter, *Eagle's Nest,* 141–43.

16. John D. Godman, *American Natural History,* 2nd ed. (Philadelphia: Key & Mielkie, 1828), 1:165, 175. The third edition was published in 1846 with Godman's writings, "Rambles of a Naturalist," included.

17. Ibid., 81–96; Thomas Sewall, *Memoir of Dr. John D. Godman* (New York: American Tract Society, 1837), 8.

18. "Dr. Godman's work though very good as far as it goes yet is inferior in respect to minute and accurate distinctions of species, as well as in not being complete," Spencer Fullerton Baird to John James Audubon, 20 June 1840, Audubon Papers, Beinecke Library; for "engraving on stone," see Charles Van Ravenswaay, *Drawn from Nature: The Botanical Art of Joseph Prestele and His Sons* (Washington: Smithsonian Institution Press, 1984), 46, and Sue W. Reed, "F. O. C. Darley's Outline Illustrations," in Ward, ed., *American Illustrated Book,* 121.

19. Bachman to J. J. Audubon, 15 March 1837, Audubon Papers, Beinecke Library.

20. John James Audubon and the Reverend John Bachman, *The Quadrupeds of North America* (New York: V. G. Audubon, 1854), 1:27; 1:349–50 on beavers; 1:269; and 2:118–23 on the opossum. This edition combines in a smaller format *The Viviparous Quadrupeds of North America* (New York: J. J. Audubon, 1845–1848), composed of imperial folio plates with accompanying text.

21. Audubon and Bachman, *The Quadrupeds*, 296–97.

22. Thomas Say, *Prospectus for American Entomology* (Philadelphia: Mitchell & Ames, 1817). Say published the prospectus and the first six plates in 1817, but reissued these plates in 1824 for *American Entomology*'s first volume; *American Entomology*, 3 vols. (Philadelphia: Samuel Augustus Mitchell, 1824–1828); Thomas Say to John F. Melsheimer, 30 July 1816, in Harry B. Weiss and Grace M. Ziegler, *Thomas Say: Early American Naturalist* (Springfield, Ill.: Charles C. Thomas, 1931), 45. The most recent Say biography is Patricia Tyson Stroud, *Thomas Say: New World Naturalist* (Philadelphia: University of Pennsylvania Press, 1992).

23. Say to Thaddeus Harris, 20 May 1830, quoted in Stroud, *Thomas Say*, 225.

24. Ibid., 224–25; Say, *Prospectus for American Entomology*.

25. Say, *American Entomology*, text for plate 12, vol. 1 (the volumes are unpaginated so that the parts eventually could be rearranged according to families and genera); text for plate 28, vol. 3; text for plate 3, vol. 1; text for plate 49, vol. 3.

26. Stroud, *Thomas Say*, 215, 227–28; Thomas Say, *American Conchology*, 3 vols. (New Harmony, Ind., 1830–1836?). See Stroud, *Thomas Say*, 266, for the work's posthumous conclusion.

27. Amos Binney, *The Terrestrial Air-Breathing Mollusks of the United States, and the Adjacent Territories of North America*, ed. Augustus A. Gould (Boston: Charles C. Little and James Brown, 1851–1857), 1:19, xxi; Charles Henry Hart, *Samuel Stehman Haldeman, LL.D.: A Memoir* (Philadelphia: [Press of E. Stern & Co.], 1881), 5–6; Samuel Stehman Haldeman, *A Monograph of the Freshwater Univalve Mollusca of the United States* (Philadelphia: Published for the Author by J. Dobson, 1842–1847).

28. Binney, *Air-Breathing Mollusks*, 1:ix; Haldeman, *Freshwater Univalve Mollusca*, part 1, 11–12 (the parts are numbered individually).

29. Haldeman, text for part 1; Norwood, *Made from This Earth*, 66–67, places Helen Lawson in a tradition of female natural history illustrators in the United States.

30. John Edwards Holbrook, *North American Herpetology*, ed. Kraig Adler ([Athens, Ohio?]: Society for the Study of Amphibians and Reptiles, 1976), retains the pagination of all five volumes of the second 1842 edition published by Judah Dobson in Philadelphia with emendations; 1:18–19.

31. Ibid., 3:35, 3:86, 4:27, 4:69, 4:132.

32. Richard D. Worthington and Patricia H. Worthington, "John Edwards Holbrook: Father of American Herpetology," in Holbrook, *North American Herpetology*, xxi–xxiii, explains the subsequently complicated history of texts and plates.

33. Asa Gray to George Engelmann, 8 April 1846, *Letters of Asa Gray*, ed. Jane Loring Gray (Boston: Houghton Mifflin, 1893), 1:340. The most complete biography of Gray remains A. Hunter Dupree, *Asa Gray, 1810–1888* (Cambridge: Belknap Press of Harvard University Press, 1959).

34. Emanual D. Rudolph, "Isaac Sprague, 'Delineator and Naturalist,'" *Journal of the History of Biology* 23 (1990); 94, 98, 101–2; Gray quote on *Genera* from Asa Gray, Letter to George Engelmann, 9 December 1846, *Letters*, 1:329; Van Ravenswaay, *Drawn from Nature*, 23–24, 45.

35. The "day wages" comment is in Asa Gray, Letter to George Engelmann, 24 January 1847, *Letters*, 1:346, and the "I could bring out . . ." quote is in Rudolph, "Isaac Sprague," 109; Ella Foshay thoroughly discusses Sprague's artistic

style in her "Nineteenth Century American Flower Painting and the Botanical Sciences," Ph.D. diss., Columbia University, 1979, 174–77.

36. Jacob Whitman Bailey to John Torrey, 6 September 1843, quoted in Michele Alexis L. Aldrich, "New York Natural History Survey," Ph.D. diss., University of Texas, 1974, 138; Ronald J. Zboray, "Antebellum Reading and the Ironies of Technological Innovation," in Davidson, *Reading in America,* 190–91, gives the figures for wages and book prices; Tyler, *Audubon's Great National Work,* 176, n. 10, calculates the octavo *Birds* cost.

37. William Starling Sullivant, quoted in Andrew Denny Rodgers III, *"Noble Fellow": William Starling Sullivant* (1940; facsimile ed., New York: Hafner, 1968), 54.

38. Tyler, *Audubon's Great National Work,* 55 and 178, n. 21, thoroughly discusses the octavo's finances; Dupree, *Asa Gray,* 168, on the demise of the *Genera;* Augustus A. Gould to James Dwight Dana, 30 September 1851, S. S. Haldeman Papers, Coll. 73, Archives, Academy of Natural Sciences, Philadelphia: "His patronage of Science, and especially his liberality in contributing this work, which must have cost him, in the neighborhood of $10,000. No copies on sales—only 290 published." Aldrich, "New York Natural History Survey," 206–7, contributes interesting figures for hand-colored engravings and lithographs (some natural history illustrations) from printers competing for the New York state survey in the early 1840s. Smallwood and Smallwood, *Natural History and the American Mind,* 183, quote a 1856 letter from James D. Dana to another zoologist stating the estimate of forty dollars for twenty-nine small wood-engraved figures.

39. Stroud, *Thomas Say,* quotes a letter from Say to Bonaparte on the publication of *Conchology,* p. 213, and another letter to Nicholas Marcellus Hentz suggesting Say's constant need for artists ("for we have not one here who can aid me as I could wish in this respect"), p. 155; Tyler, *Audubon's Great National Work,* 55–56, discusses the colorists; Van Ravenswaay, *Drawn from Nature,* 49–50.

40. Charlotte M. Porter, "The Lifework of Titian Ramsay Peale," *Proceedings of the American Philosophical Society* 129 (1985): 307; see Stroud, *Thomas Say,* 214, for Say's plans to publish reptiles; for a short biography of LeSueur, See E.-T. Hamy, *The Travels of the Naturalist Charles A. LeSueur in North America, 1815–1837,* trans. Milton Haber ([Kent, Ohio]: Kent State University Press, 1968).

41. Godman, *American Natural History,* 1:219.

42. W. Newcomb to J. J. Audubon, 17 June 1846, Audubon Papers, Beinecke Library; *Quadrupeds,* 2:63–64; Godman, *American Natural History,* 1:219, n.; Nuttall, *Manual,* 285, 276; J. Bachman to J. J. Audubon, 7 February 1846, Audubon Papers, Beinecke Library.

43. *Quadrupeds,* 1:219–20; 1:20–21; Nuttall, *Manual,* 279; Brewer, ed., *American Ornithology,* n. on pp. 4–15.

44. Brewer, ed., *American Ornithology.*

45. Holbrook, Preface to vol. 4; Say, *American Entomology,* text for plate 4, vol. 1.

46. Audubon, *Birds of America,* 6:22–31; William Oakes to J. J. Audubon, 23 January 1833, Audubon Papers, Beinecke Library; Haldeman, *Fresh Water Univalves,* pt. 5, "*Ampullaria* Genus" text.

47. Keeney in her discussion of professionalism in *The Botanizers* reminds today's student of the sciences that because of the lack of advanced degrees and career opportunities, "The terms 'amateur' and 'professional,' as commonly ap-

plied in modern usage, are simply not useful or accurate categories for the nineteenth century," p. 6. Mark Velpeau Barrow Jr., "Birds and Boundaries:
Community, Practice, and Conservation in North American Ornithology, 1865–
1935," Ph.D. diss., Harvard University, 1992, also provides a summary of scholarship concentrating on the question of professionalization, pp. 5–16.

48. See chap. 4, n. 25, for the general reading public's limited knowledge
of animal and bird species.

49. Godman, *American Natural History,* 1:129; Audubon, *Birds of America,*
6:117; Holbrook, *North American Herpetology,* 1:28, 5:9–10.

50. Audubon, *Ornithological Biography,* 5:175.

51. Holbrook, *North American Herpetology,* 3:86; *Quadrupeds,* 1:280–81; see
Nuttall on conjugal fidelity, *Manaul of Ornithology,* 9, and on the owl, 151.

52. See Nuttall, *Manual of Ornithology,* on titmice, p. 259; Audubon, *Ornitho-
logical Biography,* 2:123; Meyers, "Observations of an American Woodsman," in
Blaugrund and Stebbins, eds., *The Watercolors,* 52, and Linda Dugan Partridge,
"Domestic Violence: Scientific Themes and Audubon's Rattlesnake Attacked by
Mockingbirds" (Paper delivered at the American Studies Association Annual
Meeting, Costa Mesa, Calif., 6 November 1992), provocatively discuss intraspecies
conflict in Audubon's oeuvre.

53. Say, *American Entomology,* text for plate 16, vol. 1.

54. Brackenridge's *View of Louisiana* (1814) quoted in Graustein, *Nuttall,*
66–67; James Fenimore Cooper, *The Prairie: A Tale* (Philadelphia: Carey & Lea,
1833), 2 vols.

55. Thomas Nuttall, "Preface" to *The North American Sylva . . . Not Described
in the Work of F. Andrew Michaux* (Philadelphia: D. Rice & A. N. Hart, 1859), 1:9;
Ornithological Biography, 3:210; Elias Durand, "Biographical Notice of the Late
Thomas Nuttall," quoted in Graustein, *Nuttall,* 3.

56. Miller, "Nature and the National Ego," in *Errand into the Wilderness,*
204–16.

57. James DeKay, *Zoology of New York,* in *Natural History of New York* (Albany:
Thurlow Weed, Printer to the State, 1842), pt. 1:xii, pt. 3: 54, for synonyms for
copper-head; S[amuel] S[tehman] Haldeman, *On the Impropriety of Using Vulgar
Names in Zoology* (Philadelphia: Carey & Hart, 1843), 7; Peter J. Schmitt, *Back to
Nature: The Arcadian Myth in Urban America, 1900–1930* (New York: Oxford University Press, 1967), 34, points out that "common" names lingered into the twentieth century despite efforts to teach children the "right" names.

58. A[ugustus] A. Gould to S. S. Haldeman, 26 September 1840, Haldeman
Papers, Academy of Natural Sciences, Philadelphia; Joseph O. Pyatt, *Memoir of
Albert Newsam, Philadelphia, 1868,* 148–49, quoted in *Philadelphia: Three Centuries
of American Art* (Philadelphia: Philadelphia Museum of Art, 1976), 310; Barnhill,
"Illustrated Natural Histories," in Ward, ed. *American Illustrated Book,* 83.

59. Lucy Say, who studied with Audubon and Charles LeSueur, was an ex-
ception; for Maria Martin's background, see Alice Ford, *Audubon's Butterflies,
Moths, and Other Studies* (New York: Studio Publications, 1952), 18–19, and Lois
Barber Arnold, *Four Lives in Science* (New York: Schocken Books, 1984), 14–35.

60. Gould to Haldeman, 26 September 1840, Haldeman Papers; Alexander
Lawson Scrapbook, Coll. 79, Library, Academy of Natural Sciences; William Dun-
lap, *History of the Rise and Progress of the Arts of Design in the United States* [1834], ed.
Benjamin Blom (New York: Benjamin Blom, 1965), 3:204; J[ohn] T. Bowen to

Victor Gifford Audubon, 8 June 1848, John James Audubon Papers, Houghton Library, Harvard University; Nicholas Wainwright, *Philadelphia in the Age of Romantic Lithography* (Philadelphia: Historical Society of Pennsylvania, 1958), 46–53, 58.

61. William Swainson, *A Treatise in Malacology* (London: Longman, Orme, Brown, 1840), 264, n.

62. Erwin Stresemann, *Ornithology from Aristotle to the Present,* trans. Hans J. Epstein and Cathleen Epstein, ed. G. William Cottrell (Cambridge: Harvard University Press, 1975), remains a valuable discussion on the history of biological systematics, especially his emphasis on "exotic ornithology"; Farber, *Emergence of Ornithology,* 27–28, 106; "Christopher North" [John Wilson], "Audubon's Ornithological Biography Introduction," *Blackwood's Magazine* 30 (1831): 1–10. For an introduction to taxonomy in the twentieth century, see Ernst Mayr and Peter D. Ashlock, *Principles of Systematic Zoology,* 2nd ed. (New York: McGraw-Hill, 1991).

63. Christine Jackson, *Prideaux John Selby: A Gentleman Naturalist* (Stocksfield, Eng.: Spredden Press, 1992), 54–65; see David M. Lank, "Introductory Essay," in *Nature Classics: A Catalogue of the E. A. McIlhenny Natural History Collection at Louisiana State University,* ed. Anna H. Perrault (Baton Rouge: Friends of the LSU Library), 49, on Swainson's response to Audubon's art; Charles Wilkins Webber ["Charles Winterfield," pseud.], "American Ornithology," *American Review* 1 (1845): 271–74. The most recent biography of John Gould is Isabella Tree, *The Ruling Passion of John Gould: A Biography of the British Audubon* (London: Grove Weidenfeld, 1991).

64. David M. Knight, *Ordering the World: A History of Classifying Man* (London: Burnett Books, 1981), 93–105, 128–29; Godman, *American Natural History,* 1:xiv–xv, says, "The best system of classification in the nature of things must be in a great degree arbitrary and imperfect . . . is but a summary of distinctive epithets and characters to aid in the arrangement of knowledge"; Graustein, *Nuttall,* 228–29, quotes Nuttall's *Introduction to Systematic and Physiological Botany,* "Nature knows no rigid bounds, but plays through an infinite variety of forms, and ever avoids monotony"; Sir William Jardine, "Life of Alexander Wilson," in *American Ornithology,* ed. Sir William Jardine (London: Whittaker, Treacher, & Arnot, 1832), 1:ix–x.

65. Nuttall, *Manual of Ornithology,* 355.

66. William Oakes to John James Audubon, 23 January 1833, Audubon Papers, Beinecke Library, says, "Nuttall, you know, does not shoot"; Audubon, *Birds of America,* 1:75.

67. For example, William Cronon, "Telling Tales on Canvas: Landscapes of Frontier Change," in Jules Prown et al., *Discovered Lands, Invented Pasts: Transforming Visions of the American West* (New Haven: Yale University Art Gallery, 1992), 45–47, states that natural history illustrations such as Audubon's wild turkey (male) plate (see Illustration 5.4) indicate no habitat change; Lee Clark Mitchell, *Witnesses to a Vanishing America: The Nineteenth-Century Response* (Princeton: Princeton University Press, 1981), 1–63, and Cecelia Tichi, *New World, New Earth: Environmental Reform in American Literature from the Puritans through Whitman* (New Haven: Yale University Press, 1979), concentrate on the theme of loss in works of the major authors; Godman, *American Natural History,* 1:219.

68. *Quadrupeds,* 1:285–86.

69. Jardine, "Life of Alexander Wilson," in *American Ornithology*, 1:ix–xi.

70. See Barbara Novak, *Nature and Culture: American Landscape Painting, 1825–1875* (New York: Oxford University Press, 1980), 157–65, for an explanation of the stump's meaning; and Brian Dippie, *The Vanishing American: White Attitudes and U.S. Indian Policy* (Middletown, Conn.: Wesleyan University Press, 1982; reprint, Lawrence: University Press of Kansas, [1991]), 1–33, for the growing conventional belief that the Indian would become extinct.

CHAPTER 3

1. Quoted in Dupree, *Gray*, 211–12.

2. Quoted in Rollin H. Baker, "A Watcher of Birds," *Michigan History* 66 (September–October 1982): 42.

3. Benjamin Smith Barton, *Elements of Botany: or Outlines of the Natural History of Vegetables* (Philadelphia: Published for the author, 1803), 300–301.

4. William M. Baird and Spencer F. Baird, "List of Birds Found in the Vicinity of Carlisle, Cumberland County, Penn., about Lat. 40° 12′ N., Long. 71," *Silliman's American Journal of Science*, ser. 1, 46 (1844): 261–63.

5. For information on Kumlien, see Angie Kumlien Main, "Thure Kumlien, Koshkonong Naturalist," *Wisconsin Magazine of History* 27 (1943–1944): 17–39, 194–220, 321–43, and "Studies in Ornithology at Lake Koshkonong by Thure Kumlien," *Transactions of Wisconsin Academy of Science, Art, and Letters* 37 (1945): 91–109; J. A. A[llen], "Recent Literature," *Bulletin of the Nuttall Ornithological Club* 2 (1877): 71.

6. H[armon] A. Atkins to Edgar Alexander Mearns, 15 December 1879, quoted in Rollin H. Baker, "A Watcher of Birds," 42.

7. Edmund Berkeley and Dorothy Smith Berkeley, *A Yankee Botanist in the Carolinas: The Reverend Moses Ashley Curtis, D.D. (1808–1872)* (Berlin: J. Cramer, 1986), 51–52, 221.

8. C. W. Johnson, "Edward Sylvester Morse," *The Nautilus* 39 (April 1926): 135–36, and Dorothy G. Wayman, *Edward Sylvester Morse: A Biography* (Cambridge: Harvard University Press, 1942), 9.

9. Thomas G. Gentry, *Life-Histories of the Birds of Eastern Pennsylvania* (Philadelphia: Published by the Author, 1876), 1:xi–xiii; Berkeley and Berkeley, *Curtis*, 38–39.

10. *Illustrations of the Nests and Eggs of Birds of Ohio, with Text* with illustrations by Mrs. N. E. Jones and text by Howard Jones (Circleville, Ohio, 1879–1886), xxix, xxviii; see Ernest J. Wessen, "Jones' 'Nests and Eggs of the Birds of Ohio,' " *The Papers of the Bibliographical Society of America* 47 (Third Quarter 1953): 218–30, for a complete history of its publication. Wessen stresses that Howard E. Jones apparently relied on the observations and texts of other authors, especially his father, pp. 221–22.

11. Ibid.

12. E[lliott] C[oues], "Conclusion of the Great Work of the Nests of Ohio," *Auk* 4 (1887): 150–52; Wayman, *Morse*, 204: "The thin brown volume sold for $1.50. In the first year the sales amounted to $34.25; in 1865, to $17.00"; Main, "Studies by Thure Kumlien," 106–8. For the continuing importance of the amateur observers and collectors to the disciplines after professional careers for naturalists developed, see Marianne Ainley, "The Contribution of the Amateur to North American Ornithology," *Living Bird* 18 (1979–1980): 161–77. For the view

stressing the dwindling importance in face of discipline development and professionalization, see Keeney, *The Botanizers,* 146–50.

13. Zadock Thompson, *History of Vermont, Natural, Civil, and Statistical, in Three Parts* (Burlington, Vt.: Chauncey Goodrich, 1842), unpaginated preface, 13.

14. I[ncrease] A. Lapham, *Wisconsin: Its Geography and Topography,* 2nd ed. (Milwaukee: I. A. Hopkins, 1846), 70–74. P. R. Hoy in his obituary, "Increase A. Lapham," *Transactions of the Wisconsin Academy of Sciences, Arts, and Letters* 3 (1875–76): 265, remembered Lapham's "Herbarium of three thousand specimens—the finest in the Northwest."

15. For complete citations and history of the Massachusetts survey, see Meisel, *Bibliography,* 2:647–49; for the most complete if nonanalytical information on the natural history portions of the various surveys, see George P. Merrill, ed., *Contributions to a History of American State Geological and Natural History Surveys,* U.S. National Museum Bulletin 109 (Washington, D.C.: Government Printing Office, 1920).

16. A[ugustus] A. Gould to S. S. Haldeman, 28 January 1840, Haldeman Correspondence, Academy of Natural Sciences.

17. James E. DeKay, *Zoology of New York;* Aldrich, "New York Survey," 262, 273, 277, for numbers of described species as well as complete discussion of creation and distribution.

18. Frank Forester [Henry William Herbert], *Frank Forester's Fish and Fishing of the United States and British Provinces of North America,* 3rd ed. (New York: Stringer & Townsend, 1851), 119; William Elliott, *Carolina Sports, by Land and Water; Including Incidents of Devil-Fishing* (Charleston, S.C.: Burges & James, 1846), 55–60, calls DeKay's work "the latest authority"; *Prospectus for Natural History of the State of New York* in S. S. Haldeman Correspondence, Academy of Natural Sciences, dated 1 December 1842; John Bachman to J. J. Audubon, 15 January 1843, Audubon Papers, Beinecke Library. Smallwood and Smallwood, *Natural History and the American Mind,* 163–64, quote other naturalists' complaints about DeKay's derivative descriptions.

19. Thaddeus W. Harris, *A Treatise on Some of the Insects Injurious to Vegetation,* ed. Charles L. Flint (Boston: Crosby & Nichols, 1862), and Augustus A. Gould, *Report on the Invertebrata of Massachusetts,* 2nd ed., ed. William G. Binney (Boston: Wright & Potter, 1870); "Early State Entomologists," in Arnold Mallis, *American Entomologists* (New Brunswick, N.J.: Rutgers University Press, 1971), 37–60.

20. Captain L[orenzo] Sitgreaves, *Report of an Expedition Down the Zuni and Colorado Rivers* (Washington: Robert Armstrong, 1853), entry for 9 November [1851], 19 (there was an edition by a private printer in 1854); J. W. Abert, *Abert's New Mexico Report 1846–'47* (Albuquerque: Horn & Wallace, 1962), 8 (this is a reprint of *Report of Lieut. J. W. Abert, of His Examination of New Mexico, in the Years 1846–'47*).

21. For the instruction on collecting, see S. F. Baird, "Memoranda in Reference to the Natural History Operations," in U.S. War Department, *Reports of Explorations and Surveys to Ascertain the Most Practicable and Economic Route for a Railroad from the Mississippi River to the Pacific Ocean,* 12 vols. (Washington: A. O. P. Nicholson, 1854–1860), vol. 1, pt. 1, 9–10 (hereafter cited as *Pacific Railroad Reports*); *Lieutenant Emory Reports, a Reprint of Lieutenant W. H. Emory's "Notes of a Military Reconnaissance,"* ed. Ross Calvin (Albuquerque: University of New Mexico Press, 1951), 12–13, on Emory's botanical discoveries. John Torrey reported on

Emory's botanical specimens in W. H. Emory, *Notes of a Military Reconnaissance, from Fort Leavenworth, in Missouri, to San Diego in California* (Washington: Wendell & Van Benthuysen, 1848), 135–55; the standard works for the army surveys remain William H. Goetzmann, *Exploration and Empire* (New York: Alfred Knopf, 1966; New York: Vintage Books, 1972) and *Army Exploration in the American West, 1803–1863* (New Haven: Yale University Press, 1959). Many papers delivered at the conference "Surveying the Record: North American Scientific Exploration to 1900," American Philosophical Society Library, 14–16 March 1997, signal a fresh appreciation of the expeditions.

22. Blum, *Picturing Nature*, "Chapter Four: Scientific Prestige, National Honor: Pictures for Federal Sciences," 119–57, focuses on the U.S. Exploring Expedition, including its emulation of the French government folio productions.

23. Baird, Preface, *Pacific Railroad Reports*, vol. 8, pt. I, Mammals, xxvii; William H. Emory, *Report of the United States and Mexican Boundary Survey, Made under the Direction of the Secretary of the Interior*, 2 vols. (Washington: Cornelius Wendell, 1857–1859). The Texas State Historical Association, Austin, in 1987 published a three-volume facsimile with an introduction by William H. Goetzmann. For full explanation of the irregular (and confusing) placement of botanical and zoological sections of the Pacific Railroad reports, see Meisel, *Bibliography*, 3:100–102, Robert Taft, "The Pictorial Record of the Old West: 14—Illustrators of the Pacific Railroad Reports," *The Kansas Historical Quarterly* 19 (1951): 379–80, and John A. Moore, "The Zoology of the Pacific Railroad Survey," *American Zoologist* 26 (1986): 331–41. Taft also wrote the closely related *Artists and Illustrators of the Old West, 1850–1900* (New York: Charles Scribner's, 1952).

24. Welker, *Birds and Men*, 172–73; Woodhouse in Sitgreaves, *Expedition*, 95; Ridgway in "Ornithology," vol. 5, pt. 3, in Clarence King, *Report of the Geological Exploration of the Fortieth Parallel* (Washington: Government Print. Office, 1877), 442; Richard H. Kern to Samuel George Morton, 3 July 1850, quoted in David J. Weber, *Richard H. Kern: Expeditionary Artist in the Far Southwest, 1848–1853* (Albuquerque: Published for the Amon Carter Museum by the University of New Mexico Press, 1985), 135; H. W. Henshaw, "Ornithology," in vol. 5 of George M. Wheeler, *Report upon Geographical and Geological Explorations and Surveys West of the One Hundredth Meridian* (Washington: Government Printing Office, 1875–1889), 141.

25. For the Torrey-Schott-Prestele connection, see Brigham D. Madsen, ed., *Exploring the Great Salt Lake: The Stansbury Expedition of 1849–50* (Salt Lake City: University of Utah Press, 1989), 771–77. One letter from Schott to Torrey expresses his displeasure with Ackermann's work, 774, n 4.

26. Blum, *Picturing Nature*, extensively discusses the conventionalities involved in the federal works, pp. 165–95. Box 66, RU 7002, Spencer F. Baird Papers, 1833–1889, Smithsonian Institution Archives (hereafter cited as SIA), contains many receipts for the Mexican boundary and Pacific Railroad expeditions' drawing and printing expenses, among them a receipt from I[saac] Stevens, leader of one of the Pacific routes, to John L. LeConte for "$16.00 for eight figures of insects for report," and several to John Cassin for drawings for the Pacific Railroad expeditions. As Cassin did not draw for his own works, perhaps he supervised another draftsman in Philadelphia and disbursed the pay. For example, one 30 May 1854 receipt to William E. Hitchcock (who had worked on the Audubon octavo *Birds*), for seven drawings of birds, was paid to John Cassin.

Baird believed Richard one of the best illustrators available; he once boasted that "the drawings of snakes, fishes, etc. . . . are stupendously grand," Letter to George P. Marsh, 14 November 1854, quoted in William Healey Dall, *Spencer Fullerton Baird: A Biography* (Philadelphia: J. B. Lippincott, 1915), 313.

27. Blum, *Picturing Nature*, 161.

28. Taft gives the gigantic monetary estimate in "Pictorial Illustrators of the Pacific Railroad Reports," 380; his work, for example, and Martha A. Sandweiss, "The Public Life of Western Art," in Prown, ed., *Discovered Lands, Invented Pasts*, 119–33, focuses on the landscape views. Sandweiss, in her excerpts of criticism of the lavishly illustrated Pacific Railroad reports, notes that the scenic views and Indian portraits were considered unnecessary expenses by some, p. 128.

29. Goetzmann, *Army Exploration*, 313; "Diary," 28 June 1869, sBr 97.41.1, William Brewster Papers, Archives, Museum of Comparative Zoology Library, Harvard University.

30. Madsen, ed., *Exploring the Great Salt Lake*, 5, xii.

31. W[illiam] G[reen] Binney, *A Manual of American Land Shells*, Bulletin of the U.S. National Museum, no. 28 (Washington: Government Printing Office, 1885).

32. E. F. Rivinus and E. M. Youssef, *Spencer Baird of the Smithsonian* (Washington: Smithsonian Institution Press, 1992), 56–67, 81–97; S[pencer] F[ullerton] Baird and Charles Girard, *Catalogue of North American Reptiles in the Museum of the Smithsonian Institution* (Washington: Smithsonian Institution, 1853); Edward Lurie, *Louis Agassiz: A Life in Science* (Chicago: University of Chicago Press, 1960; rpt., Baltimore: Johns Hopkins University Press, 1988), 188–89.

33. Box 19, RU 7002, Spencer F. Baird Collection, SIA, houses the 1860s correspondence from Elliot from which the progress of his 1865 *Monograph of the Tetraoninae, or Family of the Grouse*, lithographed by Bowen & Company, may be traced.

34. Berkeley and Berkeley, *Curtis*, 151, also note that Baird funneled small amounts of money to Curtis to give to neighbors who collected; P. R. Hoy, "On the *Amblystoma Luridum*, a Salamander Inhabiting Wisconsin," in *Ninth Annual Report of the Smithsonian Institution for the Year 1854* (Washington: Smithsonian Institution, 1855), 295.

35. Preface to Baird and Girard, *Catalogue of North American Reptiles*, v–vi; Mary E. Daniels to Spencer Fullerton Baird, n.d. [1854], Incoming Correspondence of the Assistant Secretary, RU 52, box 4, folder 28 (vol. 7, 188), SIA; The Reverend Charles Mann, "On the habits of a species of Salamander, (Amblystoma opacum) Bd.," in *Ninth Annual Report*, 294; Charles Mann to Spencer Fullerton Baird, 10 March 1854, box 5, folder 11 (vol. 8, 82), Incoming Correspondence of the Assistant Secretary, RU 52, SIA; Binney, *Land Shells*, 229, 221, 22–42. For brief biographical entries on Andrews and Law, see R. Tucker Abbott, ed., *American Malacologists* (Falls Church, Va.: American Malacologists, 1973–1974), 60 (biography of George Andrews that mentions his wife), 125.

36. Lurie, *Louis Agassiz*, 188–89, notes the pleasure Agassiz's correspondents felt when such a reknowned figure replied to their questions and furthered their natural history interests; Debra Lindsay, *Science in the Subartic: Trappers, Traders, and the Smithsonian Institution* (Washington: Smithsonian Institution Press, 1993), xv, discusses the relationship with the Hudson's Bay Company and raises

the issue of motivation (e.g., the desire for increased social status and gifts of books) for Smithsonian correspondents.

37. Rachel Ann Volberg, "Constraints and Commitments in the Development of American Botany, 1880–1920," Ph.D. diss., University of California, San Francisco, 1983, explains in detail how the interests in genetics and ecology grew in one discipline. She also stresses the segmentation within the botanical community and notes that the older institutions remained associated with taxonomy and custodianship of collections.

38. Rodgers, "*Noble Fellow,*" 216.

39. Ibid., 176, 252–53, Gray quoted on 297; William Starling Sullivant, Icones Muscorum, *or Figures and Descriptions of Those Mosses Peculiar to Eastern North America Which Have Not Been Heretofore Figured* (Cambridge: Sever and Francis, 1864).

40. Carl Resek, *Lewis Henry Morgan: American Scholar* (Chicago: University of Chicago Press, 1960), 49; Aldrich, "New York State Survey," 343; Lewis H[enry] Morgan, *The American Beaver and His Works* (1868; rpt., New York: Burt Franklin, 1970), v–xi, 83, 243; one essay comparing Morgan's interests in beaver study and ethnology is Marc Sweltitz, "The Minds of Beavers and the Minds of Humans" in George W. Stocking Jr., ed., *Bones, Bodies, Behavior: Essays on Biological Anthropology* (Madison: University of Wisconsin Press, 1988), 56–83.

41. Morgan, *American Beaver,* 146–47.

42. Ibid., 225, 146, 279–84.

43. A helpful edition of the rare Cassin work is John Cassin, *Illustrations of the Birds of California, Texas, Oregon, British and Russian America*; with an introduction by Robert McCracken Peck (1856; rpt., Austin: Published for Summerless Foundation of Dallas by the Texas State Historical Association, 1991); unfortunately, no definitive biography exists for Elliot, an important transitional figure in American natural history. However, Adrian Lank in his introduction to *Pheasant Drawings by Joseph Wolf: Reproductions of the Original Sketches and the Coloured Plates of Elliot's "A Monograph of the Phasianidae, or Family of Pheasants"* (Kingston upon Hull: Allen Publishing, 1988) offers insights into the production process behind Elliot's elaborate monographs.

44. S[pencer] F[ullerton] Baird, T. M. Brewer, and Robert Ridgway, *History of North American Birds*, 3 vols. (Boston: Little, Brown & Company, 1874); Robert Ridgway, Letters to Baird, 18 August 1875, 21 October 1875, 13 August 1875, box 32, RU 7002.

45. Spencer Fullerton Baird, Letters to L[avinia] Bowen, 18 September 1875, 23 October 1875, vol. 18, microfilm reel 17, outgoing correspondence, Spencer Fullerton Baird Collection, RU 7002, SIA; L[avinia] Bowen, Letters to Baird, 27 October 1875, 1 November 1879, box 15, RU 7002. These letters evidently relate to the second edition of *History* (Little, Brown & Company, 1875), in which ironically some chromolithographs were used.

46. William H. Edwards to J[ohn] J[ames] Audubon, 14 March 1842, Audubon Papers, Beinecke Library; F. Martin Brown, ed., "The Correspondence between William Henry Edwards and Spencer Fullerton Baird," [4-part ser.], *Journal of the New York Entomological Society*, pt. 3, 67 (1958); Letter to Baird, 12 December 1867, in "Correspondence," pt. 4, 68 (1960), 166; William H. Edwards, *The Butterflies of North America*, vol. 1 (Philadelphia: American Entomological Society,

1868–1872), preface. Note that this volume is unpaginated and therefore references are made to species descriptions.

47. Reference to William G. Hewitson's "Exotic Butterflies" in Edwards to Spencer Fullerton Baird, 13 April 1863, Brown, ed., "Correspondence," pt. 3, 67 (1959), 127; Text for *Arg. Diana*, vol. 1.

48. "The Entomological Reminiscences of William Henry Edwards," *Journal of the New York Entomological Society* 59 (September 1951), 143–45, 151 (including Edwards's favorable reviews).

49. Ibid., 137, 149, 151; *Butterflies of North America*, vol. 1, text for *Parnassus* 2–4.

50. *Butterflies of North America*, text for *Grapta* 2 plate.

51. Dupree, *Gray*, 248–53.

52. Entry for 8 October [1846], *Lieutenant Emory Reports*, 89, and entry for 4 October [1846], ibid., 83; "Instructions for such as have an opportunity to Observe and Collect Cacti in the field," box 53, RU 7002, Baird Collection, SIA; George Engelmann in Emory, *Notes of a Military Reconnoissance*, 155.

53. Binney, *American Land Shells*, 18–41.

54. Willis Conner Sorenson, "Brethren of the Net: American Entomology, 1840–1880," Ph.D. diss., University of California, Davis, 1984, 279. Sorensen expanded this important work in his book of the same name (Tuscaloosa: University of Alabama Press, 1995).

55. Mark Velpeau Barrow Jr., "Birds and Boundaries: Community, Practice, and Conservation in North American Ornithology," Ph.D. diss., Harvard University, 1992, 300–318, is the fullest explanation of the developments in nomenclature. The quotation on American collections is from Alfred Newton, cited in Barrow, 309; Elliott Coues, "Fourth Installment of Ornithological Biography, Being a List of Faunal Publications Relating to British Birds," in *Proceedings of the United States National Museum* 2 (1879), 396.

56. T[homas] M[ayo] Brewer, "Some Errors Regarding the Habits of Our Birds," *American Naturalist* 1 (1867), 120; Elliott Coues, *Birds of the Colorado Valley: A Repository of Scientific and Popular Information Concerning North American Ornithology*, U.S. Geological Survey of the Territories, Miscellaneous Publications no. 11 (Washington: Government Printing Office, 1878), 600 (Wilson), 620 (Nuttall); A. A. Gould, Letter to S. S. Haldeman, 7 February 1842, Haldeman Correspondence, Library, Academy of Natural Sciences, Philadelphia.

57. Coues, *Birds of the Colorado Valley*, 166, vi. In his "The Life and Services of John James Audubon," *Transactions of the New York Academy of Sciences* 8 (1893), 53, Daniel Giraud Elliot boasts that the "new" ornithologist had evolved beyond the "woodsman" image of Audubon: "He [the ornithologist] must not only know the habits naturally and economy of birds as Audubon had, but much more . . . bibliography of subject, know Latin, French, German, Italian, anatomy and osteology for comparative relationships, even outside the specialization; pterytography, growth and structure of feathers."

58. Sorenson, "Brethren of the Net," helpfully summarizes Edwards's experiments, in the chapter "William Henry Edwards and Polymorphism in Butterflies," 310–48.

59. Baird to Edwards, 15 December 1867, Brown, ed., "Correspondence," pt. 3, 167.

60. "Zoology," vol. 5, in George M. Wheeler, *Report upon Geographical and*

Geological Exploration and Surveys West of the One Hundredth Meridian (Washington: Government Printing Office, 1875–1889), 686–92.

61. Jones, *Illustrations of the Nests and Eggs of Birds of Ohio*, xxvi–xxvii; Thure Kumlien, "On the Rapid Disappearance of Wisconsin Wild Flowers; A Contrast of the Present Time with Thirty Years Ago," *Transactions of the Wisconsin Academy of Sciences, Arts, and Letters* 3 (1876): 57.

62. Zochert, "Science and the Common Man," 463, discusses the escape from provinciality that natural history afforded to those in newly settled regions.

63. "Edward Sylvester Morse," *Nautilus* 39 (1926): 136; Harry Harris, "Robert Ridgway," *The Condor* 30 (January–February 1928): 9, reproduces Ridgway's early drawings; Julie Schimmel, "John Mix Stanley and Images of the West in Nineteenth-Century American Art," Ph.D. diss., New York University, 1983, 1–14, describes Stanley's autodidacticism; Robert V. Hine, *In the Shadow of Frémont: Edward Kern and the Art of Exploration, 1845–1860*, 2nd ed. (Norman: University of Oklahoma Press, 1982); David J. Weber, *Richard H. Kern*, 20–22.

64. See chap. 2, n. 43, for discussion of "professionalism"; Morgan, *American Beaver*, ix; Edwards, "Reminiscences," 185.

65. Rose, *Victorian America*, 8–9.

66. Scudder quotation from 1897 *Atlantic Monthly* review quoted in Edwards, "Reminiscences," 185.

67. Binney, *American Land Shells*, 18; Edward S. Morse, "Land Snails of New England," *American Naturalist* 1 (1867): 5.

68. Edwards, Letter to S[pencer] F[ullerton] Baird, 8 [November?] 1869, box 19, RU 7002, Spencer F. Baird Papers, SIA; Edwards, preface to *Butterflies*; see Elliot to Spencer Fullerton Baird, 29 December 1863; box 19, RU 7002, Spencer F. Baird Papers, SIA, on his run of two hundred copies of the grouse monograph; Baird to L[avinia] Bowen, 23 October 1875, vol. 18, 365–66, outgoing correspondence, RU 7002, SIA; Nelson Jones quoted in Wessen, "Jones' 'Birds of Ohio,' " 227.

69. The lithography firm is Tappan and Bradford, who complained in a letter to Baird, 11 September 1851, incoming correspondence of Assistant Secretary, box 1, folder 15, vol. 1, p. 346, RU 52; John H. Richard to Baird, box 32, RU 7002, SIA.

CHAPTER 4

1. Audubon Papers, Beinecke Library.

2. One study using the popular concept of "reading community" in an American historical context is Richard D. Brodhead, *Cultures of Letters: Scenes of Reading and Writing in Nineteenth-Century America* (New York: Cambridge University Press, 1993).

3. Smallwood and Smallwood, *Natural History and the American Mind*, for example, discuss the curricula and the growing lyceum lecture circuit.

4. The persons quoted on "tyrannous" plants are John Ellor Tayor and Henry Wallace Bates, in Nicolette Scourse, *The Victorians and Their Flowers* (London: Croom Helm, 1983), 59–60; Harriet Ritvo, *The Animal Estate: The English and Other Creatures in the Victorian Age* (Cambridge: Harvard University Press, 1987), focuses on class distinctions in Victorian England as expressed in practices and language concerning animals.

5. Rose, *Victorian America* 9, reinterprets in an American context M. H. Ab-

rams's explanation of romanticism's desire to cast anew the traditional relationship between God and humans as between nature and humans; Perry Miller portrays the United States as "Nature's Nation" in the chapter "Nature and the National Ego" in *Errand into the Wilderness*; his description ("Nature somehow . . . had effectively taken the place of the Bible," p. 211) remains extremely pertinent to this discussion. Turner, *Reckoning with the Beast*, discusses nature as a source of psychic comfort (p. 33) and the development of animals as a moral alternative to humankind (pp. 74–78).

6. For an overview of the antebellum American economy and its social consequences, see Charles Sellers, *The Market Revolution: Jacksonian America, 1815–1846* (New York: Oxford University Press, 1991), with its extensive bibliography; for an approach toward the flowering of antebellum culture as encouraged by the market revolution, see Anne C. Rose, *Voices of the Marketplace: American Thought and Culture, 1830–1860* (New York Twayne Publishers, 1995).

7. A recent summary of the scholarship of antebellum book publishing is Zboray, *A Fictive People*, which particularly stresses the importance of railroad lines enhancing the Northeast and the Old Northwest readership (pp. 55–67) and outlines technological innovations (p. 31). The standard works on American periodicals remain Frank Luther Mott's works, including *A History of American Magazines, 1741–1850*, vol. 1 of *A History of American Magazines* (New York: D. Appleton & Company, 1930).

8. Zboray, in "Antebellum Reading and the Ironies of Technological Innovation" in Davidson, ed., *Reading in America*, quotes publisher George Palmer Putnam's statement that in 1855 278 of the 733 works published in the United States were British imports, while 420 were original American works (pp. 180–81).

9. Peter Parley [Samuel Griswold Goodrich], *The Child's Botany*, 3rd ed. (Boston: Carter Hendee, 1830), Beinecke Library, Yale University, call no. Za G625 828Cc. The full inscription reads, "W W Atterbury / to his sweet *little* cousin— / J. M. Colt. With the hope that she will be instructed in studying the works of nature. *'Knowledge is power'* " (his italics); J[ohn] L[ee] Comstock, *Natural History of Quadrupeds* (New York: Pratt, Woodward & Co., 1848), iii.

10. *The National Union Catalog Pre-1956 Imprints* (hereafter cited as NUC), 488:558–59, lists three stereotyped editions of James Rennie's *Natural History of Birds* in the Harper's Family Library series, issued from 1840 to 1859, and his *Natural History of Insects* in the same series, issued in six editions from 1830 to 1843. The quotations are from [James Rennie], *Natural History of Birds: Their Architecture, Habits, and Faculties* (New York: Harper & Brothers, 1850), 2:106. For a discussion of the introduction of stereotyping and electrotyping into the United States, see James Exton, *The House of Harper: One Hundred and Fifty Years of Publishing* (New York: Harper & Row, 1967), 10–11, 23, and Hellmut Lehmann-Haupt, *The Book in America: A History of the Making, the Selling, and the Collecting of Books in the United States* (New York: R. R. Bowker Company, 1939), 132–34. See Patricia Anderson, *The Printed Image and the Transformation of Popular Culture, 1790–1860* (New York: Oxford University Press, 1991), for a history of publisher Charles Knight and the Society for the Diffusion of Useful Knowledge.

11. Peter Parley [Samuel Griswold Goodrich], *Peter Parley's Tales of Animals* (Boston: Carter, Hendee, & Co., 1833), "Note to Editor." The books on Goodrich remain his autobiography, *Recollections of a Lifetime*, 2 vols. (New York and Auburn,

N.Y.: Miller & Orton, 1857), and Daniel Roselle, *Samuel Griswold Goodrich, Creator of Peter Parley* (Albany: State University of New York Press, 1968), which explains Goodrich's custom of renting out not only the copyright but the actual stereotyped plates to other firms in return for a share of royalties (pp. 80–81). Although the NUC does not attribute authorship of the *Naturalist's Library* to Goodrich's publishing concern, the plates and text mirror the *Book of Ornithology* and *Tales of Animals* attributed to Peter Parley; see Sinclair Hamilton, *Early American Book Illustrators and Wood Engravers, 1670–1870* (Princeton: Princeton University Press, 1968), 1:84, entry 469; *A System of Natural History; Containing Scientific and Popular Descriptions of Various Animals* (Brattleboro, Vt.: Peck & Wood, 1834); and A[ugustus] A. Gould, ed., *The Naturalist's Library* (Boston: Phillips, Sampson, & Company, 1849).

12. S[amuel] G[riswold] Goodrich, *Johnson's Natural History; The Animal Kingdom Illustrated* (New York: A. J. Johnson & Son, 1875), 1:vi, x–xi (the Johnson publishing firm bought the stereotype plates first used in 1859 and published nine editions from 1868 to 1894, *NUC*, 440–41, under Goodrich).

13. J[ohn] L[ee] Comstock, *Readings in Zoology* (New York: Newman & Ivison, 1853), iv, 216–17, 256; Sanborn Tenney, *A Manual of Zoology*, 7th ed. (New York: Charles Scribner & Co., 1869), v–vi.

14. Margaretta Morris of Germantown, Pennsylvania, studied the Hessian fly in the 1840s. See Meisel, *Bibliography*, 3:627, for her articles; Sophia du Pont, discussed in chapter 5, copied entomological illustrations and drew insects from life. The general expectations make the contributions of Annie Law, Mary Andrews, and Valerie Blaney discussed in chapter 3 all the more impressive. For a complementary overview of Canadian women's activity in natural history during this period, see Marianne Gosztonyi Ainley, "Last in the Field?" in *Despite the Odds: Essays on Canadian Women and Science*, ed. Marianne Gosztonyi Ainley (Montreal: Vehicle Press, 1990), 25–62.

15. Almira H. Lincoln [Phelps], *Familiar Lectures on Botany*, rev. ed. (New York: Mason Brothers, 1858), 234; Keeney's *The Botanizers* richly describes nineteenth-century botanical pursuits.

16. John Darby, *Botany of the Southern States* (New York: A. S. Barnes & Co., 1855), 3.

17. Dupree, *Asa Gray*, 169, 202–3; Emanual D. Rudolph, "Almira Hart Lincoln Phelps," *American Journal of Botany* 71 (1984): 1164, n. 3, lists the Phelps printings and revisions; Berkeley and Berkeley, *Curtis*, 165, give the number for Wood's sales; see Keeney, *The Botanizers*, 12, on the significance of these sales figures.

18. See Richard Johnson, *Journal of the Society for the Bibliography of Natural History* 7 (1975): 265–67, for Say's articles on conchology; Amos Eaton, *Zoological Text-Book, Comprising Cuvier's Four Grand Divisions of Animals* (Albany, N.Y.: Webster & Skinner, 1826), v–vi; Dupree, *Gray*, 1; Rodgers, *Noble Fellow*, 110; Baird to John James Audubon, 30 January 1841, Audubon Papers, Beinecke Library.

19. Entry for 26 March 1839 in Samuel Breck, "The Diary of Samuel Breck, 1823–1840," ed. Nicholas Wainwright, *Pennsylvania Magazine of History and Biography* 103 (1979): 503; Isabelle Lehuu, "Changes in the Word: Reading Practices in Antebellum America," Ph.D. diss., Cornell University, 1992, stresses the developing divisions in readership among the classes and sexes in the midst of the general increase of printed material available. See Sally Gregory Kohlstedt, "Par-

lors, Primers, and Public Schooling: Education for Science in Nineteenth-Century America," *Isis* 81 (1990): 439, on the "noncontroversial, unifying force" of science in general.

20. Two retellings of the drake stripping the tin are in "Anecdotes of Birds," *Arthur's Home Magazine* 1 (1852): 71, and "Anecdotes of Birds," *Forester's Magazine for Youth* 10 (1852): 172–73; "A Weasel Story," *Every Youth's Gazette* 1 (1842), 14; "Natural History: Republican Swallows," Philadelphia *Saturday Courier*, 5 August 1843; "J. B.," "Benevolence in Birds—Their Usefulness, &c," *New England Farmer* 18 (1846): 280.

21. "Illustrations of Natural History," *Ballou's Pictorial Drawing-Room Companion* 8 (January–June 1855): 52–53; Harland Coultas, "The Curiosities of the Vegetable Kingdom," *Godey's Lady's Book* 51 (1855): 224–25. Zboray, *Fictive People*, 126–29, discusses the cheaper newspapers' concurrent turn to sensationalism.

22. *Gentlemen's Magazine* 22 (1752): 475; 23 (1753): 609; Isabelle Lehuu, "The 'Penny Wonder': The *New York Herald* in the Street and in the Home," in *Changes in the Word: Reading Practices in Antebellum America*, 51–92.

23. "The Canada Goose—Its Loves and Gallantry," with the subheading "Its Sagacity in Danger," *Albion*, 12 September 1840, 294; "The Blue Jay," and "The Red Thrush," [brown thrasher] *Ballou's Pictorial Drawing-Room Companion* 8 (January–June 1855), 148–49; "Bird Architecture," *Ballou's* 10 (January–July 1856), 405; [Francis C. Woodworth], "The Bottle Titmouse and Its Nest," in *The Youth's Cabinet* (New York: D. A. Woodworth, 1849) 103–4; "Perfumes for the Ladies, and Where They Come From," *Godey's Lady's Book* 51 (1855): 12; Harland Coultas, "On the Analogies between the Animal and Vegetable Kingdom," *Godey's Lady's Book* 53 (1856): 243–44.

24. "Game-Birds of America," *Graham's* 29 (1846): 249–51, 309–10; 30 (1847): 118–20, 162–64, 320–21; 31 (1847): 269–70; 32 (1848): 68–69; 33 (1848): 291–93, 358–59; Professor [John] Frost, "Wild-Birds of America," *Graham's* 34 (1849): 142, 208–9, 268, 322–24, 382–84; 35 (1849): 56–58, 126, 189–91, 246, 268, 304–5, 369; 36 (1850): 89–90. *Godey's Lady's Book* ran the pieces on fur, perfume, and shells as the lead articles for most months in volumes 50 (1855), 51 (1855), and 53 (1856); Zboray draws the connection between expansion of feeling and knowledge and women's varied reading beyond the fashions and gossip items in *A Fictive People*, "Gender and Boundlessness in Reading Patterns," pp. 156–79.

25. Henry Wetherbee Henshaw, "Autobiographical Notes," *The Condor* 21 (1919): 105; Audubon, Letter to his family, 16 [July 1842], *Journal of John James Audubon . . . 1840–1843*, ed. Howard Corning (Cambridge: Business Historical Society, 1929), 70.

26. "Ornithological Biography," *American Quarterly Review* 10 (1831); 245.

27. Christopher North [John Wilson], wrote of Audubon's appearance: "We set eyes on him in Edinburgh some five years ago . . . although he dressed somewhat after the fashion of ourselves, his long raven locks hung curling over his shoulders, from the wilderness [arranged as if] employing, perhaps, for a comb, the claw of a Bald Eagle," *Blackwood's Magazine* 30 (1831), 11; *Albion* in "J. J. Audubon, the American Ornithologist," 17 October 1840, 340, reprinted from the *London Chronicle* excerpts from the autobiographical *Ornithological Biography* material; see "Audubon, the Ornithologist," *New England Farmer* 21 (1843): 28, for the Niagara Falls hotel incident. The newspapers' eulogies upon Audubon's

travels were so well known that an amusing pastiche on his expedition to the Missouri River, "From the Rocky Mountains. Discovery of a New Animal," appears in the Philadelphia *Saturday Courier* for 19 August 1843; [Henry Tuckerman], "Birds and Audubon," *Methodist Quarterly Review* 12 (1852): 415; the admiring characterization is in G[eorge] C[heyne] Shattuck to John James Audubon, 25 July 1844, Audubon Papers, Beinecke Library.

28. Several artists portrayed Audubon with either his gun or his fur coat. The first portrait appears to be that by John Syme, done in 1828, now in the White House collection, which depicts Audubon with the hand clasping the gun to his breast. It inspired a 1834 British engraving used as a frontispiece which, in turn, inspired the image in "Audubon, the Naturalist," *Gleason's Pictorial* [which became *Ballou's*] 3 (1852): 196 (see Illustration 4.1). C. Turner in 1835 engraved a bust miniature by Frederick Cruickshank of Audubon with flowing hair, earnest expression, and an open shirt with fur collar, which Audubon gave to friends and patrons. Henry Inman's 1833 portrait, used for a popular steel engraving, also prominently shows the open shirt collar. John Woodhouse's 1843 portrait of Audubon with gun in hand and a dog at his feet inspired the Alonzo Chappel painting engraved for Evert Duyckinck's *National Portrait Gallery of Americans* (New York: Fry & Johnson, 1862). See Herrick, *Audubon the Naturalist*, 2:392–400, for further details and other portraits. Tyler, *Great National Work*, draws parallels between the portrayal of another "American woodsman," Daniel Boone, and Audubon, p. 149. John Barralet, however, may have been the first artist to associate hunting with ornithology, in his profile portrait of Alexander Wilson taken from life which shows a rifle barrel: Frank L. Burns, "Alexander Wilson: VI—Biographies, Portraits, and a Bibliography of the Various Editions of His Works," *The Wilson Bulletin*, o.s. 21 (1909): 172–74. The 1832 edition of *American Ornithology* edited by Jardine has as its frontispiece an engraving of Wilson by William Lizars from a painting by John Watson Gordon that shows Wilson clasping a gun in both hands against his chest. I wish to thank Deborah Prosser for informing me about the British conventions involved in portraying hunting gentlemen.

29. "Professor" Shelton is most likely the Reverend Shelton (d. 1853), a clergyman who was a prolific collector for John Torrey in California in the early 1850s. I am grateful to Anita L. Karg, Archivist, Hunt Institute for Botanical Documentation, for supplying biographical information on this little-known botanist, Letter to the author, 5 May 1995. A photograph in the Bancroft Library of the botanist William Brewer, ca. 1864, shows him carrying his tools around his waist and a pouch slung from his shoulders. Reproduced in Sally M. Miller, ed., *John Muir: Life and Work* (Albuquerque: University of New Mexico Press, 1993), 209.

30. Thaddeus Mason Harris, *The Natural History of the Bible* (Boston: I. Thomas & E. T. Andrews, 1793), v.

31. Jonathan Fisher, *Scripture Animals, or Natural History of the Living Creatures Named in the Bible* (Portland, Me.: William Hyde, 1834), 64–65, 182–83 (a helpful reprint was issued by the Pyne Press, Princeton, N.J., 1972, with an introduction by Mary Ellen Chase). Two biographical studies of Fisher are Chase, *Jonathan Fisher: Maine Parson, 1768–1847* (New York: Macmillan, 1948), and Alice Winchester, *Versatile Yankee: The Art of Jonathan Fisher, 1768–1847* (Princeton: Pyne Press, 1973).

32. H[enry] Harbaugh, *The Birds of the Bible* (Philadelphia: Lindsay & Blakiston, 1854), 19, 21, 25, 223–28, 258.

33. S[arah] J[osepha] Hale, *Flora's Interpreter: Or the American Book of Flowers and Sentiments*, 3rd ed. (Boston: Marsh, Capen & Lyon, 1834), v, 36.

34. Phelps, *Familiar Lectures on Botany* (1858 ed.), 155, 154, 171–72; Lucy Hooper, ed., *The Lady's Book of Flowers and Poetry* (New York: J. C. Riker, 1852), 7, 223–32; Emma C. Embury, *American Wild Flowers in Their Native Haunts* (New York: D. Appleton & Company, 1845).

35. The copy of Hooper, *Lady's Book of Flowers*, is in Eisenhower Library, Johns Hopkins University, Baltimore, call no. P8110.F6 H7 1852. Presumably the owner, Cynthia Ramage [?] of Michigan, whose name appears on the flyleaf, wrote "optimo!" above "Mrs. Balmano's Invocation to a Wreath of Transatlantic Flowers" (p. 27); Jack Goody, *The Culture of Flowers* (New York: Cambridge University Press, 1993), provides an overview of the European precedents in his chapter, "The Secret Language of Flowers in France" (pp. 232–53), and notes their continuance in the United States up to the 1900s (p. 268, n. 43), and Scourse, *Victorians and Their Flowers*, 32–38, gives the English background. Rudolph, "Isaac Sprague," 101, n. 24, points out that thousands of gift books and literary annuals published in the United States in the nineteenth century held flower plates and references.

36. Norwood, *Made from This Earth*, 12–18, sensitively discusses the blending of science and sentiment; Goody, *Culture of Flowers*, 268, also notes the American stress on morality.

37. Henry William Herbert [Frank Forester, pseud.], *The Complete Manual for Young Sportsmen* (New York: Stringer & Townsend, 1856), 283. For more background on Herbert, see David W. Judd, ed., *Life and Writings of Frank Forester (Henry William Herbert)*, 2 vols. (New York: Orange Judd Company, 1882).

38. Elisha Lewis, *Hints to Sportsmen, Containing Notes on Shooting* (Philadelphia: Lea & Blanchard, 1851), 52.

39. Henry William Herbert, *Fish of North America*, v–vi, 108, 50, 118–19.

40. Harriet Ritvo, "Learning from Animals: Natural History for Children in the Eighteenth and Nineteenth Centuries," *Children's Literature* 13 (1985): 72–90, gives the British background; *The History of Insects* (New York: Samuel Wood, 1813), 5; *The Natural History of Quadrupeds. For the Amusement of Youth* (Cooperstown, N.Y.: H. E. Phinney, 1841); *Birds of the Woodland and the Shore* (Boston: Brown, Bazin, & Co., 1855), 23 (see Illustration 5.6).

41. [Francis Lister Hawk], *Natural History; or Uncle Philip's Conversation with the Children about Tools and Trades* (New York: J. & J. Harper, 1833), 14, 38–39; Emily Taylor, *The Boy and the Birds* (New York: General Protestant Episcopal S.S. Union, 1849), 91–101.

42. Peter Parley [Samuel Griswold Goodrich], *Parley's Panorama* (Boston: Hickman, 1850), 143.

43. Comstock, *Natural History of Quadrupeds*, iii.

44. See [Thomas Bangs Thorpe], "Incidents in the Life of Audubon," *Godey's Lady's Book*, 42 (1851): 306, for "almost Christianized" quotation; J. Cypress, Jr. [H. W. Hawes], "Some Observations Concerning Quail," *American Turf Register* 11 (1840), says "I never yet heard of the parental tenderness of a trout" (p. 513).

45. Parley, *Johnson's Animal Kingdom*, ix–x; the miscellaneous notebooks in the Samuel Breck Papers, Historical Society of Pennsylvania, Philadelphia, contain "Brief Sketch of the Natural History of the Monkey Tribes," "History of the Elephant for the instruction of a school of young girls in Phila.," and "Natural history of the Cat, Composed by Samuel Breck for the amusement of the Blind

Pupils at the Institution, to whom it was read by him, in April 1854"; the sense of lessened geographical distance may correspond to the greater "annihilation of time and space," a major theme in Zboray, *A Fictive People*.

46. John B. Newman, *Boudoir Botany, Or, the Parlor Book of Flowers* (New York: Harper & Brothers, 1847), 72; Norwood, *Made from This Earth*, 40, 298, n. 72, on the encouragement of wildflower preservation.

47. *Spring Flowers; or Easy Lessons for Young Children* (Northampton, Mass.: John Metcalf, 1838), 10–11.

48. "List of Birds, found in San Francisco January 15th 1850 and offered for sale in market," J. G. Bell Diary, 1849, Document 592, Winterthur Library, notes that not only ducks but also the least sandpiper, curlew, Nootka humming-bird, bullock oriole, and Oregon snow finch were available for consumption; Samuel J. Hunter [pseud.], *The Hunters and Trappers' Illustrated Historical Guide* (St. Louis: George Knapp & Co., 1869), 158–60; the practice of caging and eating songbirds took decades to die out. Mabel Osgood Wright and Elliott Coues, *Citizen Bird, Scenes from Bird-Life in Plain English for Beginners* (New York: Macmillan, 1909), has a "Mammy" character who tells of young mockingbirds being taken from their nests to be sold in New Orleans in the recent past (p. 134); Henry Glassie in *Patterns in the Material Folk Culture of the Eastern United States* (Philadelphia: University of Pennsylvania Press, 1968), 67–74, demonstrates how baking songbirds and blackbirds in a pie persisted until the early twentieth century in some Virginia areas.

49. London *Athenaeum* review of the *Birds* dated 6 January 1838, reprinted in Waldemar Fries, *The Double Elephant Folio* (Chicago: American Libraries Association, 1973), 103; Herbert, *The Complete Manual for Young Sportsmen* (New York: Stringer & Townsend, 1857), 282–83; Dr. J[ohn] K[earsley] Mitchell, "To Audubon," 18 March 1843, Philadelphia *Saturday Courier.*

50. C[harles] W[ilkins] Webber, "The Pinneated Grouse, and the Ruffed Grouse," *Arthur's Home Magazine* 1 (1852): 81; "Ky.," "Letter to Mr. Editor," *American Turf Register* 7 (1836): 367.

51. [Original owner unknown], *Ornithological Biography* (Philadelphia: E. L. Carey and A. Hart, 1832), Library of the Massachusetts Historical Society, Gift of Charles D. Childs, 1979, vol. 1, 42; Parkman family copy of *Ornithological Biography* with margin notes, Boston Athenaeum, call no. $L9Z5 + Au2 + 3, Vol. II, 154–55. Handwriting in letters in the Audubon Papers, Beinecke Library, from George and Eliza Parkman to Audubon confirms the identities of the marginalia writers.

52. "Autobiography," in *The Letters of Asa Gray*, 1: 14–15; Angie Kumlien Main, "Thure Kumlien, Koshkonong Naturalist," *Wisconsin Magazine of History* 27 (1943–44): 333–35; *Dictionary of American Biography*, s.v. "Greene, Edward Lee"; [Original owner unknown], copy of Gray's *Botany for Young People: How Plants Grow*, Winterthur Library, call no. QK49 G77, contains pencil lines mimicking the diagrams; see Keeney, *The Botanizers*, 11–13, on the number of botanical enthusiasts.

53. Notation on flyleaf, copy of *Wilson's American Ornithology*, ed. Thomas Brewer (New York: Charles L. Cornish, 1852), Library, Academy of Natural Sciences, call no. QL681 W75. The full inscription reads: "Presented by Mrs. Elsie L. Aaron / S. F. Aaron. / Secane, Pa. / March 1917. / Procured after many years of search. this was the work with which S. F. A. first studied birds in Tennessee"; "H. A. P.," "William Healey Dall," *The Nautilus* 41 (1927): 1.

54. Harris, "Robert Ridgway," 13.

55. Frank M. Chapman, *Autobiography of a Bird-Lover* (New York: Appleton-Century, 1933), 17–18, 28–29, 31–33.

56. Maxine Benson, *Martha Maxwell: Rocky Mountain Naturalist* (Lincoln: University of Nebraska Press, 1986), 103; Walter K. Fisher, "In Memoriam: Lyman Belding," *The Condor* 20 (March–April 1918): 56–57.

57. Entry for 2 January 1856 quoted in Wayman, *Morse*, 9; Henshaw, "Autobiographical Notes," 105, and Charles Foster Batchelder, *An Account of the Nuttall Ornithological Club, 1873 to 1919*, 10–21.

58. Angie Kumlien Main, "Kumlien of Koshkonong," 323, Main, "Studies in Ornithology at Lake Koshkonong," 101; John Bachman to John James and Victor Gifford Audubon, 10 December 1845, Audubon Papers, Beinecke Library, shows Bachman asking Victor to send extracts of the German zoologist Schreiber in the original language; George C. Shattuck to John James Audubon, 15 September 1843, Audubon Papers, Beinecke Library, sent him a rare Spanish work borrowed from the Harvard Library "on [his] recognizance"; Augustus Gould checked S. S. Haldeman's reference to a French author presumably not available in Philadelphia, A. A. Gould to S. S. Haldeman, 18 October 1841, Haldeman Papers, Library, Academy of Natural Sciences; Thomas Mayo Brewer, Letter to Thure Kumlien, 25 January 1853, quoted in "Kumlien of Koshkonong," 216.

59. Zboray, "The Ironies of Technological Innovation," for book prices in comparison to wages (p. 190); Tyler, *Great National Work*, notes Thoreau's use (p. 114); Entry for 8 September 1841, "Journal May 1841–November 1843," Spencer Fullerton Baird Papers, Box 41, RU 7002, Smithsonian Institution Archives.

60. "H.A.P.," "William Henry Dall," 1; Henshaw, "Autobiographical Notes," 105; William Brewster in his "The Black-and-Yellow Warbler (Dendroeca Maculosa)," *Bulletin of the Nuttall Ornithological Club* 2 (1877), 4, emotes in Audubonian fashion in an attempt to engage the reader: "Be careful how you shake that branch, for I would have you take a good long look ere we disturb her. See how her dark little eye glistens, and note the rapid pulsating motion of her back. Underneath those puffed-up fathers a poor little heart is beating wildly with fear and apprehension, but still she sits still on her trust."

61. Margaret French Cresson, *Journey into Fame: The Life of Daniel Chester French* (Cambridge: Harvard University Press, 1947), 27–28; William French to Daniel Chester French, 29 April 1866, Daniel Chester French Papers, Microfilm Roll 39, Manuscript Division, Library of Congress, shows the brothers' joint interest in taxidermy. William despairs of close friend and future ornithologist Ruthven Deane's skill while secure in his own ability: "worse and worse instead of better, for the last that he stuffed (a female cow bunting that I think is a very easy bird), is a frightful looking creature with a neck that looks about half as long as its body, but he seems almost to think it perfection."

62. The Harris speech is quoted in "Audubon's Ornithology, First Volume," *American Journal of Science*, 1st ser., 39 (1840): 349; William Francis Brand, *Life of William Rollinson Whittingham* (New York: E. & J. B. Young, 1883), 1: 5, 7, 14–15.

63. Thomas Wentworth Higginson, *Cheerful Yesterdays* (1899; reprint New York: Arno Press and the New York Times, 1968), 24–25, 26, 88–89.

64. "A Lady" [Susan Fenimore Cooper], *Rural Hours*, 3rd ed. (New York: George Putnam, 1851), 10, 86, 377, 404; Susan Fenimore Cooper, Letter to Victor Gifford Audubon, 28 October 1851, Audubon Papers, Beinecke Library, thanks him: "In looking over the article on the Moose I find that it will be of

very essential service to me." Norwood, *Made from This Earth*, 25–41, thoroughly analyzes the content and structure of *Rural Hours*.

65. Joseph A. Moore, Letter to John James Audubon, 28 September 1842, Audubon Papers, Beinecke Library, on the pet crow, "Buck," whom he knew in his childhood; M. V. B. Morrison, *The Orphan's Experience: Or the Hunter and Trapper* (Des Moines: Steam Printing House of Mills Iowa, 1868), 173.

66. Entry for 29 July 1837, John Collins Warren, Journal, vol. 1, Warren Papers, Library, Massachusetts Historical Society.

67. Entries for 27 March, 4 April, 27 April [1807], Diary of Henry Moore, Queen Anne County Library (photocopy at Manuscripts Division, Maryland Historical Society). I would like to thank Donna Ellis, former manuscripts librarian, for pointing out this particular diary's interest in botany.

68. Entry for 5 April 1855, *Diary of Caroline Seabury, 1854–1863*, ed. Suzanne L. Bunkers (Madison: University of Wisconsin Press, 1991), 39; Lewis O. Saum, *The Popular Mood of Pre–Civil War America* (Westport, Conn.: Greenwood Press, 1980), 176–77, 196, has noted the "surprising number of people" trying to describe landscape as sublime.

69. Entries for 27 June, September 20, 1853, Caleb B. R. Kennerly Diary 1853–1854, MS. #1926, Manuscript Division, Library of Congress.

70. Entries for 7 April, 26 August, James G. Cooper, "General Notes—1853," James G. Cooper Papers, 1853–1870, RU 7067, SIA; Entries for 25 May 1853, July 4 1853, "Daily journal of George Suckley While serving on the Pac. RR Survey, May 24 to Dec. 6, 1853," George Suckley Papers, 1849–61, RU 7191, SIA; for the Hammond anecdote, see [George Suckley], "Report on Mammals for Northwest Boundary Survey [1857–61]," 5, George Suckley Papers, RU 7191.

71. Anonymous, "C & O Canal Journal 1858," Manuscript Division, Library of Congress, 8–9, 60–62 [the cover on the journal is 1858, but the date given on the first page of the text is 1859]. This memoir is published in Ella E. Clark and Thomas F. Hahn, eds., *Life on the Chesapeake and Ohio Canal 1859* (Shephardstown, W.Va.: American Canal and Transportation Center, 1975), but lacks the illustrations. I wish to thank the reading room staff of the LC Manuscript Division for their help in bringing this and other travel writings to my notice.

72. Ibid., 46; the writer also implied that belief in illogical explanations crossed racial lines: he mocks "an otherwise intelligent man in New Hampshire who declared that the common 'rose bug' lost its wing covers [to become a horse fly]" (p. 47).

CHAPTER 5

1. Doris M. Bowman, *The Smithsonian Treasury: American Quilts* (Washington: Smithsonian Institution Press, 1991), 48; Virginia Eisemon, Textiles Collection, National Museum of American History, Letter to author, 18 November 1996.

2. Patricia Anderson summarizes concerns expressed by authors such as Roland Barthes, Walter Benjamin, John Berger, and William Ivins about changes created by such "recontextualizations" in *The Printed Image and the Transformation of Popular Culture*, p. 5.

3. Hamilton, *Early American Book Illustrators*, remains the most complete source on antebellum wood engraving, although relatively few periodical images are discussed. In an increasing number of instances, the Research Libraries Infor-

mation Network (RLIN) holds information on individual engravers, such as Abel Bowen, who signed the illustrations in works not indicated in Hamilton's book.

4. Peter J. Parker and Stefanie Munsing Winkelbaum, "Embellishments for Practical Repositories: Eighteenth-Century American Magazine Illustration," in *Eighteenth-Century Prints in Colonial America,* ed. Joan D. Dolmetsch (Charlottesville: The Colonial Williamsburg Foundation, distributed by the University Press of Virginia, 1979), 71–97; Noah Webster, *History of Animals: Designed for the Instruction and Amusement of Persons of Both Sexes* (New Haven: Howe & Deforest and Walter & Steele, 1812), copy in Rare Books Department, Library of Congress, QL50/ .W38.

5. Frances Hinklin, *Bewick Wood Engravings* (London: Her Majesty's Stationery Office, 1978), 15–16, 18–19, records Bewick's own observation that one of his wood-engraved vignettes served over 900,000 printings; Hinklin states that Bewick's blocks still print well. For an introduction to Bewick, see Iain Bain, *Thomas Bewick: An Illustrated Record of His Life and Work* (Newcastle-upon-Tyne: Laing Gallery, 1979); Lawrence Thompson, *Alexander Anderson: His Tribute to the Wood-Engraving of Thomas Bewick* (Princeton: Princeton University Press, 1940).

6. For further explanations of stereotyping and electrotyping in relation to illustration, see Lehmann-Haupt, *The Book in America,* 132–33, and Bamber Gascoigne, *How to Identify Prints* (New York: Thames & Hudson, 1986), sections 71 and 72; John Wright, ed., *Buffon's Natural History of the Globe and of Man* (London, 1831), 4 vols.; *A Natural History of the Globe, of Man, of Beasts, Birds, Fishes, Reptiles, Insects and Plants* (Boston: Gray & Bowen, and Philadelphia: Thomas Desilver, 1831) 1:iii–iv; 4:165, 178; [Edward Turner Bennett], *The Tower Menagerie: Comprising the Natural History of the Animals Contained in That Establishment* (London, 1829) and *Gardens and Menageries of the Zoological Society* (London: Thomas Tegg & N. Hailes, 1830–1831).

7. François André Michaux, *The North American Sylva,* trans. Augustus L. Hillhouse (Paris: Printed by C. d'Hautel, sold by various American publishers, 1818), 3 vols. (This is probably the edition that was most available for the Parley concern's copying.) *A Pictorial Geography of the World* [Sinclair Hamilton, #470], 62–69, has the Michaux *North American Sylva* tree adaptations; Peter Parley [Samuel Griswold Goodrich], *Peter Parley's Tales of Animals* (Boston: Carter, Hendee, & Co., 1833); *Goldsmith's Natural History: Abridged for the Use of Schools by Mrs. Pilkington, Revised and Corrected, by a Teacher of Philadelphia* (Philadelphia: M. Polock, 1857), 92, reproduces the same jaguar as *The Naturalist's Library,* ed. A. A. Gould (Boston: Phillips, Sampson, & Company, 1849) (see Illustration 5.3); Peter Parley [Samuel Griswold Goodrich], *Book of the United States, Geographical, Political, and Historical* (Boston: Charles J. Hendee, 1837), 102–8, runs the same bird images of the Audubonian wild turkey, the Wilsonian pinneated grouse, and the Bennett/Harvey wild swan found in *A Book of Ornithology* (Boston: William Hyde & Co., 1832). In one of many examples of the prevalence of these Parley/Bennett/Harvey wood engravings, "The Male and Female Ostrich" on p. 1 of *Youth's Gazette* 1, no. 5 (26 February 1842), published in New York, is probably from the same block used in Parley's *Manual of Ornithology* and *Naturalist's Library.* See chapter 4, n. 9, for Goodrich's practice of renting out stereotyped plates.

8. For background on Bowen, see William Henry Whitmore, "Abel Bowen," *Collections of the Bostonian Society* (Boston, 1887), no. 2, and Hamilton, 1:78–80, 81–83. Unfortunately, William James Linton in the important *History of Wood-*

Engraving in America (London: George Bell & Sons, 1882), 16, dismisses Bowen as "simply a mere copier" but does reproduce an image from Nuttall's *Ornithology*. Comstock, *Readings in Zoology,* reproduces many of the *Manual of Ornithology* engravings; *The Child's Story Book of Birds* (Macon, Ga.: J. W. Burke, 1867).

9. [Francis Lister Hawk], *Natural History,* 208, shows the Audubonian oriole nest; Tenney, *A Manual of Zoology,* v–vii; Mrs. Sanborn [Abby Amy] Tenney, *Pictures and Stories of Animals for the Little Ones at Home* (New York: Sheldon & Company, 1870), contains six volumes in the series: Quadrupeds; Birds; Fishes and Reptiles; Bees, Butterflies, and Other Insects; Sea Shells and River Shells; Sea-Urchins, Star-Fishes, and Corals.

10. For example, in her 1858 edition of *Familiar Lectures,* the May apple (fig. 168, p. 246) adapted Gray's *Genera,* vol. 1, plate 35, of the same plant.

11. For the exact whippoorwill image (placed upside-down in relation to the Audubon original), Professor [John] Frost, "Wild-Birds of America," *Graham's* 34 (1849): 208; this same image accompanies a poem by Elizabeth F. Ellett in *Arthur's Ladies Magazine* 2 (1844), 224; *Godey's Lady's Book* 53 (November 1856): 395 and *Ballou's* 9 (July–December 1855): 180 used the same block or duplicate blocks to produce the shell imagery; the engraved lines match exactly. Several of the birds in "Natural History—Illustrations of Ornithology," *Ballou's* 10 (1856), 64, resemble images in Charles Knight, *The Pictorial Kingdom* [London, 1844], and wood engravings throughout the Society for the Diffusion of Useful Knowledge's *Penny Magazine* in the 1830s and 1840s published in London and available in the United States. Indeed, the preface in the first volume of the American edition of Knight's and the Society's *Penny Magazine* (1832) states that stereotypes cast from the original London blocks were used. Gascoigne, *How to Identify Prints,* section 71, also notes that stereotype blocks from English periodicals were exported to "no less than seventeen countries" in 1837.

12. See Henry C. Watson, *Thrilling Adventures of Hunters in the Old World and the New* (Boston: Kelley & Brothers, 1853), 327, for Bewick "bison" adaptation; John Frost, *Pictorial Family Encyclopedia of History, Biography, and Travels* (Buffalo, N.Y.: Miller, Orton & Mulligan, 1854), 186, 188, for its two Bewick "bisons" and 244 for its "Parley" bears; Samuel J. Hunter [pseud.], *The Hunters' and Trappers' Illustrated Guide,* shows the "Parley" owl and swan (pp. 63 and 180); Frank Forester [Henry William Herbert], *Frank Forester's Field Sports of the United States and British Provinces of North America,* 6th ed. (New York: Stringer & Townsend, 1851), and "Advertisement" in preface material for *Fish of the United States.*

13. Inscription from *Juvenile History of Birds. Part 1* (New York: Mahlon Day & Co., n.d.), Winterthur Library, Rare Book PZ6/ J97h; a childish outline in ink mimics the horse in a piece of wallpaper pasted into the back cover of a drawing book, Benjamin H. Coe, *Easy Lessons in Landscape Drawing* (Hartford, Conn.: E. B. & E. C. Kellogg, 1842), Winterthur Library, Rare Book NC615 /C67 s.

14. Benson, *Maxwell,* 2; C[harles] W[ilkins] Webber, *Wild Scenes and Song-Birds* (New York: George P. Putnam & Co., 1854), 75; Harris, "Ridgway," 13; Chapman, *Autobiography of a Bird-Lover,* 17.

15. T. H. Clarke, *The Rhinoceros from Dürer to Stubbs, 1515–1790* (London: Philip Wilson for Sotheby's Publications, 1986); see also n. 2 above for changes in context.

16. Images of many big cats and other exotic animals possibly originating in the British *Tower Menargarie* and *Zoological Gardens* in 1829 and 1831 reappeared

in a 1834 Barre, Massachusetts, newspaper advertisement promoting a menagerie, probably adapted from intermediary Parley wood engravings (*Farmer's Gazette,* September 1834, no. 16, p. 3). I wish to thank Myron Stachiw for this reference.

17. Mott, *A History of American Magazines, 1741–1850,* 552. Gleason's peak circulation figures were about 100,000, Mott, *A History of American Magazines, 1850–1865* (1939; rpt., Cambridge: Belknap Press of Harvard University Press), vol. 2 of *A History of American Magazines,* 411; "Bird Architecture," *Ballou's Pictorial,* 10 (1856): 405; Lewis, *Hints,* 2nd ed., xiv–xv; Tenney, *Manual of Zoology,* viii. See chap. 2, n. 35, for references including illustration production costs.

18. For an extended discussion of Gosse's and the general British interest in microscopic and marine life imagery, see Lynn L. Merrill, *The Romance of Victorian Natural History* (New York: Oxford University Press, 1989), 107–38, 190–214. For the beginning of American interest in microscopy in the 1850s, see John Hailey Warner, " 'Exploring the Inner Labyrinth of Creation': Popular Microscopy in Nineteenth-Century America," *Journal of the History of Medicine and Allied Sciences* 37 (January 1982):7–33.

19. I would like to thank the helpful staff of the Lynn Historical Society for researching the dates of the Pike's establishment. Advertising Card Album no. 26, 70 × 130.26, 15–18, Joseph Downs Collection of Manuscripts and Printed Ephemera, Winterthur Library, holds examples of the Church & Co.–sponsored sets of Audubon-derived trade cards. The Warshaw Collection of Business Americana, Archives Center, National Museum of American History, Washington, holds a rich and varied collection of ornithologically derived trade cards, mostly under the subheading "Birds" in Collection 60. Many cards printed abbreviated life histories with the species' geographical distribution and habits (often stressing the economic advantages of songbirds to agriculture) on the verso; this same collection, Box 1, "Animals," folder 25, holds proofs and the chromolithographed illustrations based on figures in Audubon's *Quadrupeds* for an unknown publication. Unfortunately, Robert Jay, *The Trade Card in Nineteenth-Century America* (Columbia: University of Missouri Press, 1987), does not discuss natural history in its history of trade card production and use for advertising purposes; Robert McCracken Peck, *A Celebration of Birds: The Life and Art of Louis Agassiz Fuertes* (New York: Walker & Company, 1982), 22, 46, 89.

20. For example, Benjamin Nutting, *Nutting's Picture Sketches* (Boston, 1862), plate 13, draws the outlines of a generalized songbird; *Drawing for Young Children* (New York: Charles S. Francis, 1853) shows squirrels, pigs, and a pink. Peter C. Marzio, *The Art Crusade: An Analysis of American Drawing Manuals, 1820–1860,* Smithsonian Studies in History and Technology, no. 34 (Washington: Smithsonian Institution Press, 1976), 77, however, does list one example: *The Alphabetical Drawing-Book, and Pictorial Natural History of Quadrupeds* (New York: Wiley & Putnam, 1847). Gavin R. Bridson and James J. White, *Plant, Animal and Anatomical Illustration in Art and Science: A Bibliographical Guide from the Sixteenth Century to the Present Day* (Winchester, Eng.: St. Paul's Bibliographies, in assoc. with Hunt Institute for Botanical Documentation, 1990), lists the European manuals devoted to flower and animal painting but notes the small amount of material devoted to teaching scientific illustration, xvi.

21. Amelia Peck, *American Quilts and Coverlets in the Metropolitan Museum of Art* (New York: The Metropolitan Museum of Art and Dutton Studio Books,

1990), 154, explains that a "Mrs. Hicks" may have contributed the wool that professional weavers (probably male, hence "craftsmen") used to make the coverlet. The cat form in the coverlet closely resembles the wood engraving of "The Chetah [*sic*], or Hunting Leopard" in the *Penny Magazine* (25 January 1834), 32.

22. Peck, *American Quilts*, 233, describes the coverlet.

23. "American Entomology," *North American Review*, n.s. 23 (July 1825), 251; Webber, "The Quadrupeds of North America," *The American Review* 4 (1846): 625–26; *Silliman's Journal of the Arts and Sciences* 39, no. 2 (1840): 393 on Haldeman; *North American Review* quoted in Rodgers, *Noble Fellow*, 252; *Science* for 9 October 1885, quoted in W. H. Edwards, "Reminiscences," 159.

24. Barnhill, "Illustrated Natural Histories," in Ward, ed., *American Illustrated Book*, 83, on Holbrook's lithotints; John Fiske Allen and William Sharp, *Victoria Regia* (Boston: Printed and published for the Author, by Dutton & Wentworth, 1854).

25. Fries, *Double Elephant Folio*, 151–71, lists the American subscribers with helpful descriptions; "List of Subscribers," *Viviparous Quadrupeds of North America* (New York: J. J. Audubon, 1846–1851), vols. 1 and 2; authors have estimated the number of first edition octavo *Birds* sold at over 1,200, but Ron Tyler, *Great National Work*, has reduced the numbers to 1,000 or 1,050 (p. 101); see Herrick, *Audubon the Naturalist*, 2:391, "Note by the Author," for a count of octavo *Quadrupeds* subscribers; Tyler, *Great National Work*, studies the later productions, 112–25, 129, 164, 185, 192.

26. See Margaret Welch, "John James Audubon and His American Audience: Art, Science, and Nature, 1830–1860," Ph.D. diss., University of Pennsylvania, 1988, 151–56, for discussion of sources of wealth. The work of Edward Pessen, including his *Riches, Class, and Power: America before the Civil War* (Lexington, Mass: D. C. Heath, 1973), remains helpful in its general portrayal of the antebellum wealthy. The Boston subscribers and their milieu have drawn the most sustained attention in articles such as Paul Goodman, "Ethics and Enterprise: The Values of a Boston Elite, 1800–1860," *American Quarterly* 18 (1966): 435–51, Robert Dalzell, *Enterprising Elite: The Boston Associates and the World They Made* (Cambridge: Harvard University Press, 1987), and Betty G. Farrell, *Elite Families: Class and Power in Nineteenth-Century Boston* (Albany: State University of New York Press, 1993).

27. See Welch, "John James Audubon," for a fuller explication of artistic and natural history tastes that was derived from wills, inventories, the Boston Athenaeum's book circulation records, and art exhibition records (pp. 185–222).

28. Audubon and Bachman, *Quadrupeds*, 1:90; Majorie B. Cohn, *Francis Calley Gray and Art Collecting for America* (Cambridge: Harvard University Art Museums, 1986), quotes the viewing pleasure of Jonathan Mason Warren, a neighbor of the Bostonian subscribers (p. 239). Both Cohn and Georgia Brady Baumgardner, "Print Collecting in Antebellum America," in *From Artist to Patron: The Fraser Collection of Engravings Presented to Dr. Robert Gibbes* (Columbia, S.C.: McKissick Museum and the Institution for Southern Studies, [1985]), 25–33, explicate this period's use of prints for home decoration as well as serious collecting.

29. See Elizabeth Johns, *American Genre Painting: The Politics of Everyday Life* (New Haven: Yale University Press, 1991), 22–23, on New York City's status as an economic and social center; Herrick, *Audubon the Naturalist*, 2: 292–93, on octavo *Quadrupeds* sales; Annette Blaugrund, "The Artist as Entrepreneur," in *John James*

Audubon, ed. Blaugrund and Stebbins, on the folio *Birds* promotion (pp. 30–35); Tyler, *The Great National Work,* 69–71, on Audubon's extremely successful door-to-door sales of the octavo *Birds.*

30. Photograph of Dundas Library in Dundas-Lippincott Mansion, Frederic Gutekunst, c. 1875, P. 8704.2, Library Company of Philadelphia; *Diary of George Templeton Strong,* ed. Allan Nevins (New York: Macmillan, 1952), 1:344; Charles Frederick Beck, Will 87, Year 1859, County of Philadelphia; Richard W. Morin, "Statesman and Artist," *Dartmouth College Library Bulletin,* n.s., 10 (1969): 7–9; I am grateful to Deborah Dependahl Waters, Curator for Decorative Arts, Museum of the City of New York, Letter to the author, 8 June 1994, for describing the Jenny Lind cabinet and books with custom bindings.

31. Entry for 26 January 1849, *Diary of George Templeton Strong,* 1:344; Fries, *Double Elephant Folio,* 274, quotes Captain Frederick Marryat's 1838 observation about the Library of Congress ("I saw a copy of Audubon's Ornithology, and many other works, in a very dilapidated state; but this must be the case when the library is open to all"); William Oakes to John James Audubon, 18 March 1833, Audubon Papers, Beinecke Library.

32. George Washington Doane, "The Word of God to be Studied with His Works: The Introductory Lecture before the Burlington Lyceum," in *George Washington Doane, Bishop of New Jersey,* ed. William Doane (New York, 1860), 4:399.

33. The series of letters from the English subscriber John Heppenstall to Audubon's engraver, Robert Havell, demonstrates that Heppenstall owned or was thoroughly familiar with the artwork of natural history illustrators Jacques Barraband and Thomas Bewick, 30 November 1831, 1 March 1837, Audubon Papers; Joseph A. Moore to John James Audubon, 7 March 1846, Audubon Papers, Beinecke Library.

34. See nn. 26 and 27 above.

35. C[harles] W[ilkins] Webber, "The Quadrupeds of North America," *The American Review* 4 (1846): 636–37.

36. G[eorge] C[heyne] Shattuck to John James Audubon, 25 July 1844, Audubon Papers, Beinecke Library; R. B. Rhett to John James Audubon, 4 February 1841, Audubon Papers, Beinecke Library; Elizabeth Fothergill to John James Audubon, 1 March 1842, Audubon Papers, Beinecke Library.

37. Johns, *American Genre Painting,* 1–23; Albert Boime, *The Magisterial Gaze,* Angela Miller, *The Empire of the Eye,* and the work of Alan Wallach, including "Making a Picture of the View from Mount Holyoke," *American Iconology,* 80–91, which discusses Daniel Wadsworth, link the panoramic landscape's "proprietary" viewpoint to its patrons.

38. I borrow the phrases about the "other" from Todorov, *The Conquest of America: The Question of the Other,* 247; see chapter 2, n. 43, for Daniel Wadsworth reference; Fries, *Double Elephant Folio,* 244–45, gives the excerpt from a committee report communicated by Oscar W. Sturtevant to the Board of Assistant Aldermen of the City of New York concerning the purchase of the *Birds* (the resolution passed 26 February 1850).

39. Tyler, *Great National Work,* 82, 103–4; Robert McCracken Peck, in his introduction to John Cassin's *Illustrations of the Birds of California,* notes that Cassin admitted that he would personally have to canvass subscribers in order to make the production a financial success (pp. 1–21).

40. Entry for 11 December [1841], "Journal May 1841–November 1843," Spencer Fullerton Baird Papers, Box 41, RU 7002, SIA, notes "copying Fringilla

Iliaca at Mr. Audubon's," and several oversize drawings in the same collection show Baird copying popular drawing books, diagrams of animal heads, and Audubon plates (dated 20 and 23 December 1841). Winchester, *Versatile Yankee*, reproduces many examples of Fisher's copies and life studies, including "A careful imitation of a Carrot raised in Bluehill, in the year 1807, in the garden of the Rev. Jonathan Fisher—Drawn and Painted by candlelight Jan. 28 1808. By Jon. Fisher, Bluehill."

41. Betty-Bright Low and Jacqueline Hinsley, eds., *Sophie du Pont, A Young Lady in America: Sketches, Diaries, & Letters, 1823–1833* (New York: Harry N. Abrams, 1987), discuss Sophie's natural history interests (she once attempted to copy Wilson's *Ornithology*) and reproduce many of her entomological drawings (pp. 184–88).

42. Norman B. Wilkinson, *E. I. du Pont, Botaniste: The Beginning of a Tradition* (Charlottesville: University Press of Virginia, Published for the Eleutherian Mills-Hagley Foundation, 1972), 82–83, reproduces Victorine's notes on plant anatomy taken from botanical manuals; Entry for 25 June 1832, Diary–1832, 70, Samuel Breck Papers, Historical Society of Pennsylvania, mentions, "in my walk in the woods, I meet for many weeks the above little flowers" (unfortunately the watercolors he painted are now missing from the diary).

43. Certainly academic artists other than Tait and Hays were influenced by the natural history disciplines, as Ella Foshay demonstrates in her classic article on Martin Johnson Heade, "Charles Darwin and the Development of American Flower Imagery," *Winterthur Portfolio* 15 (1980): 299–314. Many scholars have noted Thomas Cole's and Frederick Church's interest in geology. However, I believe that the nonacademic artists more clearly demonstrate intersections with natural history discourse; another painting of various species of birds perching on branches reveals varying understanding of current ornithological illustration. An unknown figure, "T. R.," entitled an oil on wood image to a friend, relative, or sweetheart, Martha Jane: "Humming Bird./ Red Bird./ Baltimore Bird./ Robbin./ Flicker./ Blue Bird, Spring 1842." The labeling defines the specific species, although not in scientific names. The birds' clearly positioned feet, eyes with a large highlight, and the vigorous twist of the cardinal's head in the upper right corner relate the painter's style to the ornithological conventions in contemporary literature. The artist has enlisted these aspects to create a pleasing image to give as a keepsake. Richard J. Wattenmaker and Alain G. Joyaux, *American Naive Paintings: The Edgar William and Bernice Chrysler Garbisch Collection* (Flint, Mich.: Flint Institute of Arts, 1981), 72–73; Deborah Chotner, *American Naive Paintings* (Washington: National Gallery of Art and Cambridge University Press, 1992), 421–23, provides helpful information on the *Birds* identifications; Susan C. Waters, who was an itinerant painter on the New York–Pennsylvania border in the 1840s, had begun her art career by "painting copies for the course in Natural History" in exchange for her and her sister's tuition at Friendsville (Pa.) Boarding School for Females in the 1830s. ("Painting copies" meant producing posterlike enlargements for classroom use.) In her portraits scholars have noted "botanically accurate plants" in pots, as in her 1845 Henry C. Wells portrait, in which she "imagines" the green plant with forms not conventionally pretty but derived from the distinctive forms of a Venus flytrap. See Colleen Cowles Heslip, "Susan C. Water," *Antiques* 115 (1979): 767–77, and Chotner, *American Naive Paintings*, 388–91.

44. Diary of Susan Nichols, Joseph Downs Collection of Manuscripts and

Printed Ephemera, Winterthur Library, No. 65 × 607; Album and Scrapbook of Eleuthera du Pont Smith, Joseph Downs Collection of Manuscripts and Printed Ephemera, Winterthur Library, 65 × 623.1, 9 [butterfly], 89 [moth image].

45. "To Make Pictures of Birds with Their Natural Feathers," *Godey's Lady's Magazine* 50 (1855): 129–30; J. H. Mintorn, *Lessons in Flower and Fruit Modelling in Wax* (London: George Routledge & Sons, n.d.); and "The Art of Making Wax Fruits and Flowers," *Godey's Lady's Magazine* 53 (1856): 421–23.

46. Joseph H. Batty, *Practical Taxidermy, and Home Decoration* (New York: Orange Judd Company, 1880); C. J. Maynard, *The Naturalist's Guide in Collecting and Preserving Objects of Natural History* (Boston: Fields, Osgood, & Co., 1870), 36, bids his readers: "Therefore I say to those who would learn to mount birds in natural attitudes, study nature. Have all attitudes that every bird assumes engraved upon the brain, . . . from the one assumed by the delicate Warbler, that hops lightly from limb to limb, . . . to that of the mighty Eagle in his eager, downward swoop upon his trembling prey." A 1882 letter from Spencer Fullerton Baird remarks that the owl population was decreasing owing to their popularity "for ladies headgear and as ornaments for the parlor and library" (quoted in Rivinus and Yossef, *Baird*), 112; Main, "Studies in Ornithology at Lake Koshkonong," 101.

47. Bowman, *The Smithsonian Treasury: American Quilts,* 48.

48. A fine example of a zoological quilt is reproduced and discussed in Susanna Pfeffer, *Quilt Masterpieces* (New York: Park Lane, 1990), fig. 24. Two figures are derived from Audubon's wild turkey (Illustration 5.4), and one deer figure strongly resembles Bewick's stag (Illustration 5.2).

49. Marsha MacDowell and Ruth D. Fitzgerald, eds., *Michigan Quilts: 150 Years of a Textile Tradition* (East Lansing: Michigan State University Museum, 1987), 40, and Kathleen Common Schmidt, telephone conversation with author, 20 February 1997. Mrs. Schmidt, the quilter's granddaughter, relates that Kimball owned history and geographical textbooks and possibly subscribed to *Godey's Lady's Book*. Kimball also expressed a love of nature in letters and quotations of poetry she copied. The literature about historic quilts using figurative designs is significant, with Pat Ferrero and Elaine Hedges, *Hearts and Hands: The Influence of Women and Quilts on American Society* (San Francisco: Quilt Digest Press, 1987), Patsy Orlofsky and Myron Orlofsky, *Quilts in America* (1974; reprint, New York: Abbeville, 1992), and Jane Bentley Kolter, *Forget Me Not: A Gallery of Friendship and Album Quilts* (Pittstown, N.J.: Main Street Press, 1985), serving as useful introductions. Jacqueline M. Atkins and Phyllis Tepper, *New York Beauties: Quilts from the Empire State* (New York: Dutton Studio Books in association with the Museum of American Folk Art, 1992), discuss the variety produced in New York state but do not reproduce any quilts directly affected by natural history. Figurative embroidery also portrayed shells, birds, and flowers, with Mrs. E. G. Miner of Canton, New York, contributing the most spectacular example of a natural history adaptation in her embroidered carpet, exhibited in 1844. It copied in remarkable detail the ruffed grouse in Wilson's *American Ornithology* (the quilt is reproduced in *Antiques* 9 [1926]: 400).

50. H. B. Möllhausen, sketch labeled "Opuntia Engelmanni var. cyclodes / Gallinas river / Sept 25.1853," botanical specimen in the Amiel Whipple Collection, Archives and Manuscripts Division of the Oklahoma Historical Society. George Engelmann notes that his artist, Paulus Roetter, made the final drawing for this cactus species, but "The drawings made on the spot by Mr. H. B. Möll-

hausen, the artist of the expedition, greatly aided the work and were made use of, and even partly copied," in his "Description of the Cactaceae," *Pacific Railroad Reports,* 4:57–58.

51. I am indebted to Amy Meyers's discussion of Peale's "imposit[ion of] traditional structures of knowledge upon all that he saw," in her chapter "The Accessible Environment: Specimen Drawings and Landscapes by Titian Ramsay Peale," in *Sketches from the Wilderness,* 194–258. Kenneth Haltman, "Private Impressions and Public Views: Titian Ramsay Peale's Sketchbooks from the Long Expedition, 1819–1820," *Yale University Art Gallery Bulletin* 40 (1989), also reproduces the image (p. 38). Haltman also stresses the "guided visual access provided through compositional techniques" of natural history and ethnographical illustration (p. 42). Blum, *Picturing Nature,* reproduces a more finished version of this squirrel in the same lively attitude from another sketchbook (p. 93).

52. Greenfield Village and Henry Ford Museum, *Lewis Miller's Guide to Central Park* (Dearborn, Mich.: Edison Institute, 1977); for accounts of Miller's life and works, see Donald A. Shelley, *Lewis Miller: Sketches and Chronicles* (York, Pa.: The Historical Society of York County, 1966).

53. Eugene C. Worman Jr., "The Watercolors and Prints of Orra White Hitchcock," *AB Bookman* (13 February 1989), 646–55.

54. Beatrice Scheer Smith, *A Painted Herbarium: The Life and Art of Emily Hitchcock Terry (1838–1921)* (Minneapolis: University of Minnesota Press, 1992), 6–7, 29–32, 184.

Conclusion

1. John Burroughs, *Wake-Robin,* 5th ed. (Boston: Houghton Mifflin, 1890), 262.

2. Barrow, "Birds and Boundaries," 22–62, examines the implications of postwar economic development for natural history magazine publishing and the rising taxidermy industry.

3. "Answers to Circulars: Smithsonian Correspondents, Subjects in which interested," Box 64, RU 7002, Spencer Fullerton Baird Papers, SIA.

4. [Phelps], *Familiar Lectures on Botany,* 235.

5. "Answers to Circulars"; the individual survey forms are filed alphabetically by surname.

6. Ibid.

7. Ibid.; see Barrow, "Birds and Boundaries," pp. 33–35, for a definition of "entrepreneurial natural history" and pp. 28–31 for the growing science museum movement.

8. "Answers to Circulars."

9. Elliott Coues, *Key to North American Birds,* 5th ed. (Boston: Page Company [1927]); the definitive biography on Coues is Paul Russell Cutright and Michael J. Brodhead, *Elliott Coues: Naturalist and Frontier Historian* (Urbana: University of Illinois Press, 1981); Joseph Kastner, *A World of Watchers* (New York: Alfred A. Knopf, 1986), 50, points out the need for the bird specimen in Coues's process; A. S. Packard Jr., *Guide to the Study of Insects, and a Treatise on Those Injurious and Beneficial to Crops* (Salem, Mass.: Naturalist's Book Agency, 1869), was another introductory manual. For an introduction to Peterson's career, see John C. Devlin and Grace Naismith, *The World of Roger Tory Peterson: An Authorized Biography* (New York: Times Books, 1977). See Felton Gibbons and Deborah Stone, *Neighbors to*

the Birds: A History of Birdwatching in America (New York: W. W. Norton, 1988), 287–309, for a history of the twentieth-century bird guides.

10. John Henry Comstock, *Insect Life: An Introduction to Nature-Study* (New York: D. Appleton, 1903), 88; Peter Schmitt's still influential *Back to Nature: The Arcadian Myth in Urban America, 1900–1930* (New York: Oxford University Press, 1969), 86–95, discusses "Nature Study" in the classroom from 1900 to 1940; Keeney, *The Botanizers,* 137–45.

11. Comstock, *Insect Life,* 87.

12. See *The American Naturalist* 1 (1872) for both E. S. Morse, "The Land Snails of New England," 5–16, and Elliott Coues, "The Quadrupeds of Arizona," 281–92. A recent edition of a 1939 work, *Life Histories of North American Woodpeckers,* with original paintings by William Zimmerman (Bloomington: Indiana University Press, 1992), is a good example of Bent's works.

13. Randall Lockwood, "Anthropomorphism Is Not a Four-Letter Word," in *Perceptions of Animals in American Culture,* ed. R. J. Hoage (Washington: Smithsonian Institution Press, 1989), 41–56, discusses "Morgan's Canon" and the historically negative perception of anthropomorphism in the scientific community. For more detail, see C. Lloyd Morgan, *Animal Life and Intelligence* (Boston: Ginn & Company, 1891), in which he stresses the human capacity for abstract thought, and so asks, "How far does the dog construct a similar world?" (p. 336).

14. See Lisa Mighetto, "Science, Sentiment, and Anxiety: American Nature Writing at the Turn of the Century," *Pacific Historical Review* 45 (1985): 33–50, and *Wild Animals and American Environment Ethics* (Tucson: University of Arizona Press, 1991), 1–25, for important discussions of the popular literature. Ralph H. Lutts, *The Nature Fakers: Wildlife, Science, and Sentiment* (Golden, Colo.: Fulcrum, 1990), fully chronicles the early twentieth-century controversy over some popular natural history literature deemed untruthful by some naturalists such as John Burroughs and Theodore Roosevelt; Mabel Osgood Wright and Elliott Coues, *Citizen Bird, Scenes from Bird-Life in Plain English for Beginners,* 240–42.

15. Burroughs, *Wake-Robin,* 260, 263. One recent biography is Edward Rehenan, *John Burroughs: An American Naturalist* (Post Mills, Vt.: Chelsea Green, 1992).

16. Maurice Thompson, *Sylvan Secrets, in Bird-Songs and Books* (New York: John B. Alden, 1887), 10, 7; Thomas R. Dunlap, *Saving America's Wildlife* (Princeton: Princeton University Press, 1988), 18–33, analyzes this generation's need to reconcile Romanticism and Darwinism in its attitude toward animals; the scholarship on the American nature essay is abundant, stressing the profound influence of Henry David Thoreau on later generations. For an introduction see Robert Finch and John Elder, *The Norton Book of Nature Writing* (New York: W. W. Norton, 1990), Frank Stewart, *A Natural History of Nature Writing* (Washington: Island Press, 1995), and John Cooley, ed., *Earthly Words: Essays on Contemporary American Nature and Environmental Writers* (Ann Arbor: University of Michigan Press, 1994). James McClintock, " 'Pray without Ceasing': Annie Dillard among the Nature Writers," in *Earthly Words,* 69–70, remarks, "Nature writing in America has always been religious or quasi-religious" despite suspicions about traditional religion.

17. Peck, *A Celebration of Birds,* 29, on Fuertes's life, reproduces a remarkably "Audubonian" 1895 watercolor drawing; Nicholas Hammond, *Twentieth Century Wildlife Artists* (Woodstock, N.Y.: Overlook Press, 1986), 59–63, discusses Eckel-

berry. William V. Mealy and Peter Friederici, eds., *Value in American Wildlife Art: Proceedings of the 1992 Forum* (Jamestown, N.Y.: Roger Tory Peterson Institute of Natural History, 1992), contains many essays on past and present-day animal artists, including Robert A. McCabe, "American Wildlife Art: Roots, Influences, and Similarities," 22–29, which explicitly traces today's artists' stated influences; Roger Tory Peterson, "Bird Painting in America," *Audubon Magazine* 44 (May–June 1942), 22.

18. Allen Rokach and Anne Millman, *Focus on Flowers: Discovering and Photographing Beauty in Gardens and Wild Places* (New York: Abbeville Press, 1990), 39, comments on the popularity of the "straightforward portrait" of a flower species; Michael Freeman, *The Wildlife and Nature Photographer's Field Guide* (Cincinnati: Writers' Digest Books, 1984), 18, 62, 64, notes that despite the interest in the relationships among species, "Most wildlife photography . . . normally includes a single main subject."

19. On Muir's botanical efforts, see Nancy M. Slack, "Botanical Exploration of California from Menzies to Muir (1786–1900)," in *John Muir, Life and Work*, ed. Sally M. Miller, 196–201, 234–36; Paul Russell Cutright, *Theodore Roosevelt the Naturalist* (New York: Harper & Brothers, 1956), 2; the George Bird Grinnell quotation is from his article in *Forest and Stream*, "The Audubon Society," 11 February 1886, reprinted in John F. Reiger, *American Sportsmen and the Origins of Conservation*, rev. ed. (Norman: University of Oklahoma Press, 1986), 67–69. For a history of the Audubon Society, see Frank Graham with Carl W. Bucheister, *The Audubon Ark: A History of the National Audubon Society* (New York: Alfred A. Knopf, 1990). Stephen R. Fox, *John Muir and His Legacy: The American Conservation Movement* (Little, Brown, 1981), notes the important activities of conservation groups throughout the twentieth century.

20. Barrow, "Birds and Boundaries," traces the often uneasy relationship between ornithologists and other leaders in the bird protection movement, 383–416, 430–525. T. S. Palmer, *Legislation for the Protection of Birds Other Than Game Birds*, United States Bureau of Biological Survey, Bulletin 12, rev. ed. (Washington, 1902), remains the most complete source for nineteenth-century bird protection legislation; Robin W. Doughty, *Feather Fashions and Bird Preservation: A Study in Nature Protection* (Berkeley: University of California Press, 1975), 101, notes that the first *Audubon Magazine,* designed to impart the message of respect for birds, ran articles about Audubon's and Wilson's lives in the late 1880s, thereby emphasizing the tradition of studying and empathizing with birds.

21. A model for tracing the split between amateurs and professional scientists in a specific discipline is Keeney, *The Botanizers*.

22. Spencer, Fullerton Baird, Letter to George P. Marsh, 2 May 1852, quoted in Rivinus and Youssef, *Baird*, 92. Rivinus and Youssef called this directive "possibly the most unusual instruction ever levied on a minister plenipotentiary of the United States"; Kastner, *World of Watchers*, 156–68, discusses the women writers and educators in the early twentieth century.

23. For example, Douglas, *Feminization American Culture*, in her preface, "The Legacy of American Victorianism" (pp. 3–13), draws parallels between nineteenth-century sensibilities and twentieth-century popular culture.

Select Bibliography

The following bibliography lists only sources cited more than once in the notes and captions.

Primary Sources: Printed and Manuscript

Abert, J. W. *Abert's New Mexico Report 1846-'47*. Albuquerque: Horn & Wallace, 1962.

Audubon, John James. *The Birds of America*. London, 1827–1838.

———. *The Birds of America*. Philadelphia, 1840–1844. Reprint, New York: Dover, 1967.

———. Papers. General MSS. 85. Beinecke Rare Book and Manuscript Library, Yale University, New Haven, Conn.

Audubon, John James, and Rev. John Bachman. *The Quadrupeds of North America*. 3 vols. New York: V. G. Audubon, 1854.

———. *The Viviparous Quadrupeds of North America*. 2 vols. New York: Published by J. J. Audubon, 1845–1848.

Baird, Spencer F[ullerton]. Incoming Correspondence of the Assistant Secretary, 1850–1877. RU 52. Smithsonian Institution Archives, Washington, D.C.

———. Papers, 1833–1889. RU 7002. Smithsonian Institution Archives, Washington, D.C.

Baird, S[pencer] F[ullerton], T. M. Brewer, and Robert Ridgway. *History of North American Birds*. 3 vols. Boston: Little, Brown & Company, 1874.

Baird, S[pencer] F[ullerton], and Charles Girard. *Catalogue of North American Reptiles in the Museum of the Smithsonian Institution*. Washington: Smithsonian Institution, 1853.

Bartram, William. *Travels through North and South Carolina, Georgia, East and West Florida*. London, 1792. Facsimile ed., Charlottesville: University of Virginia Press, by arrangement with the Beehive Press, 1980.

Batty, Joseph H. *Practical Taxidermy, and Home Decoration*. New York: Orange Judd Company, 1880.

Belknap, Jeremy. *History of New-Hampshire*. 3 vols. Boston: Bradford & Read, 1818.

Binney, Amos. *The Terrestrial Air-Breathing Mollusks of the United States, and the Adjacent Territories of North America*. Ed. Augustus A. Gould. Boston: Charles C. Little and James Brown, 1851–1857.

Binney, W[illiam] G[reen]. *A Manual of American Land Shells*. Bulletin of the U.S. National Museum, no. 28. Washington: Government Printing Office, 1885.

Birds of the Woodland and the Shore. Boston: Brown, Bazin & Company, 1855.

Breck, Samuel. Papers. Historical Society of Pennsylvania, Philadelphia.

Brown, F. Martin, ed. "The Correspondence between William Henry Edwards and Spencer Fullerton Baird." Parts 1, 3, 4. *Journal of the New York Entomological Society* 66 (1958): 191–223; 67 (1959): 125–49; 68 (1960): 157–75.

Burroughs, John. *Wake-Robin.* 5th ed. Boston: Houghton Mifflin, 1890.

"C&O Canal Journal 1858." Manuscript Division. Library of Congress, Washington, D.C.

Cassin, John. *Illustrations of the Birds of California, Texas, Oregon, British and Russian America.* Philadelphia: J. B. Lippincott, 1856. Reprint, with an introduction by Robert McCracken Peck, Austin: Published for the Summerless Foundation of Dallas by the Texas State Historical Association, 1991.

Catesby, Mark. *The Natural History of Carolina, Florida, and the Bahama Islands.* London, 1731–43.

Chapman, Frank M. *Autobiography of a Bird-Lover.* New York: D. Appleton-Century, 1933.

Comstock, Anna Botsford. *The Comstocks of Cornell: John Henry Comstock and Anna Botsford Comstock.* Edited by Glenn W. Herrick and Ruby Green Smith. Ithaca, N.Y.: Comstock Publishing Associates, 1953.

Comstock, J[ohn] L[ee]. *Natural History of Quadrupeds.* New York: Pratt, Woodward & Co., 1848.

Cooper, James Fenimore. *The Prairie: A Tale.* 2 vols. Philadelphia: Carey & Lea, 1833.

Cooper, James G. Papers, 1853–1870. RU 7067. Smithsonian Institution Archives, Washington, D.C.

Coues, Elliott. *Birds of the Colorado Valley: A Repository of Scientific and Popular Information Concerning North American Ornithology.* U.S. Geological Survey of the Territories, Misc. Publications, no. 11. Washington: Government Printing Office, 1878.

DeKay, James. *Zoology of New York.* 5 vols. In *Natural History of New York.* Albany: Thurlow Weed, Printer to the State, 1842–1844.

Drayton, John. *A View of South-Carolina as Respects Her Natural and Civil Concerns.* Charleston, S.C.: W.P. Young, 1802.

Edwards, William H. *The Butterflies of North America.* Vol. 1. Philadelphia: American Entomological Society, 1868–1872.

Edwards, William Henry. "The Entomological Reminiscences of William Henry Edwards." Edited by Cyril F. dos Passos. *Journal of the New York Entomological Society* 59 (1951): 129–86.

Emory, William H. *Notes of a Military Reconnaissance, From Fort Leavenworth, in Missouri, to San Diego in California.* Washington: Wendell & Van Benthuysen, 1848.

———. *Report of the United States and Mexican Boundary Survey, Made Under the Direction of the Secretary of the Interior.* 2 vols. Washington: Cornelius Wendell, 1857–1859.

Forester, Frank [Henry William Herbert]. *Frank Forester's Fish and Fishing of the United States and British Provinces of North America.* 3rd ed. New York: Stringer & Townsend, 1851.

Frost, Professor [John]. "Wild-Birds of America." *Graham's* 34 (1849): 142, 208–9, 267–68, 322–24, 382–84; 35 (1849): 56–58, 126, 189–91, 246, 268, 304–5, 369; 36 (1850): 89–90.

[Gerard, John]. *The Herball or General Historie of Plantes. Gathered by John Gerard . . . Very Much Enlarged and Amended by Thomas Johnson.* London, 1633.

Godman, John D. *American Natural History*. 2nd ed. Philadelphia: Key & Mielkie, 1828.

Gould, Augustus A. *A Report on the Invertebrata of Massachusetts*. 2nd ed. Edited by William G. Binney. Boston: Wright & Potter, 1870.

Gray, Asa. *Letters of Asa Gray*. Edited by Jane Loring Gray. 2 vols. Boston: Houghton Mifflin, 1893.

Haldeman, Samuel Stehman. *A Monograph of the Freshwater Univalve Mollusca of the United States*. Philadelphia: Published for the Author by J. Dobson, 1842–1847.

Haldeman, S[amuel] S[tehman]. Papers. MSS. Collection 73. Library, Academy of Natural Sciences, Philadelphia.

[Hawk, Francis Lister.] *Natural History; or Uncle Philip's Conversations with the Children about Tools and Trades*. New York: J. & J. Harper, 1833.

Holbrook, John Edwards. *North American Herpetology*. Edited by Kraig Adler. Philadelphia, 1842. Reprint, [Athens, Ohio?]: Society for the Study of Amphibians and Reptiles, 1976.

Hunter, Samuel J. [pseud.] *The Hunters' and Trappers' Illustrated Historical Guide*. St. Louis: George Knapp & Co., 1869.

Illustrations of the Nests and Eggs of Birds of Ohio, with Text. With illustrations by Mrs. N. E. Jones and text by Howard Jones. Circleville, Ohio: n.p., 1879–1886.

King, Clarence. *Report of the Geological Exploration of the Fortieth Parallel*. Washington: Government Printing Office, 1877.

Lawson, Alexander. Scrapbook. MSS. Collection 79. Library, Academy of Natural Sciences, Philadelphia.

Lawson, John. *A New Voyage to Carolina*. Edited by Hugh Talmage Lefler. 1709. Reprint, Chapel Hill: University of North Carolina Press. 1967.

Madsen, Brigham D., ed. *Exploring the Great Salt Lake: The Stansbury Expedition of 1849–50*. Salt Lake City: University of Utah Press, 1989.

Miller, Lewis. *Lewis Miller's Guide to Central Park*. Dearborn, Mich.: Edison Institute, 1977.

Morgan, Lewis H[enry]. *The American Beaver and His Works*. 1868. Reprint, New York: Burt Franklin, 1970.

A Natural History of the Globe, of Man, of Beasts, Birds, Fishes, Reptiles, Insects and Plants. Boston: Gray & Bowen, and Philadelphia: Thomas Desilver, 1831.

Nichols, Susan. Diary. No. 65 × 607. Joseph Downs Collection of Manuscripts and Printed Ephemera, Winterthur Library, Winterthur, Del.

Nuttall, Thomas. *A Manual of the Ornithology of the United States and of Canada*. 2 vols. Cambridge, Mass.: Hilliard & Brown, 1832, 1834.

"Ornithological Biography." *American Quarterly Review* 20 (1831): 245–58.

"Ornithological Biography." *Loudon's Magazine of Natural History* 8 (1835): 184–90.

Parley, Peter. [Goodrich, Samuel Griswold]. *Book of the United States, Geographical, Political & Historical*. Boston: Charles J. Hendee, 1837.

———. *Johnson's Natural History: The Animal Kingdom Illustrated*. New York: A. J. Johnson and Son, 1875.

———. *Peter Parley's Tales of Animals*. Boston: Carter, Hendee, & Co., 1833.

[Phelps], Almira H. Lincoln. *Familiar Lectures on Botany*. Rev. ed. New York: Mason Brothers, 1858.

Say, Thomas. *American Entomology*. 3 vols. Philadelphia: Samuel Augustus Mitchell, 1824–1828.

Sitgreaves, Captain L[orenzo]. *Report of an Expedition Down the Zuni and Colorado Rivers*. Washington: Robert Armstrong, 1853.

Smith, Eleuthera du Pont. Album and Scrapbook. No. 65 × 623.1. Joseph Downs Collection of Manuscripts and Printed Ephemera, Winterthur Library, Winterthur, Del.

[Smithsonian Institution]. *Ninth Annual Report of the Smithsonian Institution, for the Year 1854*. Washington: Smithsonian Institution, 1855.

Tenney, Sanborn. *A Manual of Zoology*. 7th ed. New York: Charles Scribner & Co., 1869.

U.S. War Department. *Reports of Explorations and Surveys to Ascertain the Most Practicable and Economic Route for a Railroad from the Mississippi River to the Pacific Ocean*. 12 vols. Washington: A. O. P. Nicholson, 1854–1860.

Webber, Charles Wilkins [Charles Winterfield, pseud.]. "American Ornithology." *American Review* 1 (1845): 262–74.

———. "The Quadrupeds of North America." *The American Review* 4 (1846): 625–38.

Wheeler, George M. *Report upon Geographical and Geological Explorations and Surveys West of the One Hundredth Meridian*. Washington: Government Printing Office, 1875–1889.

Williams, Samuel. *Natural and Civil History of Vermont*. 2nd ed. Burlington, Vt.: Printed by Samuel Mills, 1809.

Wilson, Alexander. *American Ornithology*. 9 vols. Philadelphia: Bradford & Inskeep, 1808–1814.

———. *The Life and Letters of Alexander Wilson*. Edited by Clark Hunter. Memoirs of the American Philosophical Society, vol. 154. Philadelphia: American Philosophical Society, 1983.

———. *Wilson's American Ornithology: With Additions Including the Birds Described by Audubon, Bonaparte, Nuttall, and Richardson*. Edited by Thomas M. Brewer. 1840. Reprint, New York: Arno Press, 1970.

Wright, Mabel Osgood, and Elliott Coues. *Citizen Bird, Scenes from Bird-Life in Plain English for Beginners*. New York: Macmillan, 1909.

Secondary Sources

Aldrich, Michele Alexis L. "New York Natural History Survey." Ph.D. diss., University of Texas, 1974.

Allen, David Elliston. *The Naturalist in Britain: A Social History*. London: Allen Lane, 1976.

Anderson, Patricia. *The Printed Image and the Transformation of Popular Culture, 1790–1860*. New York: Oxford University Press, 1991.

Arber, Ann. *Herbals: Their Origin and Evolution*. 2nd ed. London: Cambridge University Press, 1938.

Baker, Rollin H. "A Watcher of Birds." *Michigan History* 66 (September–October 1982): 40–43.

Barrow, Mark Velpeau, Jr. "Birds and Boundaries: Community, Practice, and Conservation in North American Ornithology, 1865–1935." Ph.D. diss., Harvard University, 1992.

Batchelder, Charles Foster. *An Account of the Nuttall Ornithological Club, 1873 to*

1919. Memoirs of the Nuttall Ornithological Club, no. 8. Cambridge: The Club, 1937.

Benson, Maxine. *Martha Maxwell: Rocky Mountain Naturalist.* Lincoln: University of Nebraska Press, 1986.

Bercaw, Nancy Dunlap. "Solid Objects/Mutable Meanings: Fancywork and the Construction of Bourgeois Culture, 1840–1880." *Winterthur Portfolio* 26 (Winter 1991): 232–47.

Berkeley, Edmund, and Dorothy Smith Berkeley. *A Yankee Botanist in the Carolinas: The Reverend Moses Ashley Curtis, D.D. (1808–1872).* Berlin: J. Cramer, 1986.

Blaugrund, Annette, and Theodore E. Stebbins Jr., eds. *The Watercolors for* The Birds of America. New York: Villard Books and the New-York Historical Society, 1993.

Blum, Ann Shelby. *Picturing Nature: American Nineteenth-Century Zoological Illustration.* Princeton: Princeton University Press, 1993.

Boime, Albert. *The Magisterial Gaze: Manifest Destiny and American Landscape Painting, c. 1830–1865.* Washington: Smithsonian Institution Press, 1991.

Bowman, Doris M. *The Smithsonian Treasury: American Quilts.* Washington: Smithsonian Institution Press, 1991.

Brewster, William. *October Farm: From the Concord Journals and Diaries of William Brewster.* Cambridge: Harvard University Press, 1936.

Brown, Richard D. *Knowledge Is Power: The Diffusion of Information in Early America, 1700–1865.* New York: Oxford University Press, 1989.

Cantwell, Robert. *Alexander Wilson: Pioneer and Ornithologist.* Philadelphia, Lippincott, 1961.

Chotner, Deborah. *American Naive Paintings.* Washington: National Gallery of Art and Cambridge University Press, 1992.

Dall, William Healey. *Spencer Fullerton Baird: A Biography.* Philadelphia: J. B. Lippincott, 1915.

Dance, S. Peter. *The Art of Natural History: Animal Illustrators and Their Work.* Woodstock, N.Y.: Overlook Press, 1978.

Daniels, George H. *American Science in the Age of Jackson.* New York: Columbia University Press, 1968.

Davidson, Cathy N., ed. *Reading in America: Literature and Social History.* Baltimore: Johns Hopkins University Press, 1989.

Douglas, Ann. *The Feminization of American Culture.* New York: Alfred A. Knopf, 1977.

Dupree, A. Hunter. *Asa Gray, 1810–1888.* Cambridge: Belknap Press of Harvard University Press, 1959.

Ellenius, Allan, ed. *The Natural Sciences and the Arts: Aspects of Interaction from the Renaissance to the 20th Century.* Uppsala: Almquist & Wiksell International, 1985.

Farber, Paul Lawrence. *The Emergence of Ornithology as a Scientific Discipline, 1760–1850.* Dordrecht, Neth.: D. Reidel, 1982.

Ford, Alice. *John James Audubon: A Biography.* New York: Abbeville Press, 1988.

Franklin, Wayne. *Discoverers, Explorers, and Settlers: The Diligent Writers of Early America.* Chicago: University of Chicago Press, 1979.

Fries, Waldemar H. *The Double Elephant Folio.* Chicago: American Library Association, 1973.

Fritzell, Peter A. *Nature Writing and America: Essays upon a Cultural Type.* Ames: Iowa State University Press, 1990.

Gascoigne, Bamber. *How to Identify Prints.* New York: Thames & Hudson, 1986.

Goetzmann, William H. *Army Exploration in the American West, 1803–1863.* New Haven: Yale University Press, 1959.

Goody, Jack. *The Culture of Flowers.* New York: Cambridge University Press, 1993.

Graustein, Jeannette E. *Thomas Nuttall, Naturalist: Explorations in America, 1808–1841.* Cambridge: Harvard University Press, 1967.

Greene, John C. *American Science in the Age of Jefferson.* Ames: Iowa State University Press, 1984.

Hamilton, Sinclair. *Early American Book Illustrators and Wood Engravers, 1670–1870.* Princeton: Princeton University Press, 1968.

Harris, Harry. "Robert Ridgway." *The Condor* 30 (January–February 1928): 5–118.

Herrick, Francis Hobart. *Audubon the Naturalist.* 2nd ed. 1938. Reprint, New York: Dover, 1968.

Johns, Elizabeth. *American Genre Painting: The Politics of Everyday Life.* New Haven: Yale University Press, 1991.

Kastner, Joseph. *A World of Watchers.* New York: Alfred A. Knopf, 1986.

Keeney, Elizabeth B. *The Botanizers: Amateur Scientists in Nineteenth-Century America.* Chapel Hill: University of North Carolina Press, 1992.

Lehmann-Haupt, Hellmut. *The Book in America: A History of the Making, the Selling, and the Collecting of Books in the United States.* New York: R. R. Bowker, 1939.

Lehuu, Isabelle. "Changes in the Word: Reading Practices in Antebellum America." Ph.D. diss., Cornell University, 1992.

Lurie, Edward. *Louis Agassiz: A Life in Science.* 1960. Reprint, Baltimore: Johns Hopkins University Press, 1988.

Main, Angie Kumlien. "Studies in Ornithology at Lake Koshkonong by Thure Kumlien." *Transactions of Wisconsin Academy of Science, Art, and Letters* 37 (1945): 91–109.

———. "Thure Kumlien, Koshkonong Naturalist." *Wisconsin Magazine of History* 27 (1943–44): 17–39, 194–220, 321–43.

McNamara, Kevin R. "The Feathered Scribe: The Discourses of American Ornithology before 1800." *William and Mary Quarterly* 47 (1990): 210–34.

Meisel, Max. *A Bibliography of American Natural History: The Pioneer Century, 1769–1865.* 3 vols. Brooklyn, N.Y.: Premier Publishing, 1924.

Meyers, Amy R. Weinstein. "Sketches from the Wilderness: Changing Conceptions of Nature in American Natural History Illustration, 1680–1880." Ph.D. diss., Yale University, 1985.

Miller, Angela. *The Empire of the Eye: Landscape Representation and American Cultural Politics, 1825–1875.* Ithaca, N.Y.: Cornell University Press, 1993.

Miller, David C., ed. *American Iconology: New Approaches to Nineteenth-Century Art and Literature.* New Haven: Yale University Press, 1993.

Miller, Perry. *Errand into the Wilderness.* Cambridge: Belknap Press of Harvard University Press, 1956.

Miller, Sally M. *John Muir, Life and Work.* Albuquerque: University of New Mexico Press, 1993.

Mott, Frank Luther. *A History of American Magazines.* 5 vols. Cambridge: Harvard University Press, 1930–1968.

Norwood, Vera. *Made from This Earth: American Women and Nature.* Chapel Hill: University of North Carolina, 1993.

Novak, Barbara. *Nature and Culture: American Landscape Painting, 1825–1875.* New York: Oxford University Press, 1980.

Peck, Amelia. *American Quilts and Coverlets in the Metropolitan Museum of Art.* New York: The Metropolitan Museum of Art and Dutton Studio Books, 1990.

Peck, Robert McCracken. *A Celebration of Birds: The Life and Art of Louis Agassiz Fuertes.* New York: Walker & Company, 1982.

Porter, Charlotte. *The Eagle's Nest: Natural History and American Ideas, 1812–1842.* Tuscaloosa: University of Alabama Press, 1986.

Prown, Jules, et al. *Discovered Lands, Invented Pasts: Transforming Visions of the American West.* New Haven: Yale University Art Gallery, 1992.

Regis, Pamela. *Describing Early America: Bartram, Jefferson, Crèvecoeur, and the Rhetoric of Natural History.* De Kalb: Northern Illinois University Press, 1992.

Rivinus, E. F., and E. M. Youssef. *Spencer Baird of the Smithsonian.* Washington: Smithsonian Institution Press, 1992.

Rodgers, Andrew Denny, III. *"Noble Fellow": William Starling Sullivant.* 1940. Reprint, New York: Hafner, 1968.

Rose, Anne C. *Victorian America and the Civil War.* New York: Cambridge University Press, 1992.

Rudolph, Emanuel D. "Isaac Sprague, 'Delinator and Naturalist.' " *Journal of the History of Biology* 23 (1990): 91–126.

Scourse, Nicolette. *The Victorians and Their Flowers.* London: Croom Helm, 1983.

Smallwood, William Martin, and Mabel Smallwood. *Natural History and the American Mind.* New York: Columbia University Press, 1941.

Smith, Beatrice Scheer. *A Painted Herbarium: The Life and Art of Emily Hitchcock Terry (1838–1921).* Minneapolis: University of Minnesota Press, 1992.

Sorenson, Willis Conner. "Brethen of the Net: American Entomology, 1840–1880." Ph.D. diss., University of California, Davis, 1984.

Stroud, Patricia Tyson. *Thomas Say: New World Naturalist.* Philadelphia: University of Pennsylvania Press, 1992.

Taft, Robert. *Artists and Illustrators of the Old West, 1850–1900.* New York: Charles Scribner's Sons, 1953.

———. "The Pictorial Record of the Old West: 14: Illustrators of the Pacific Railroad Reports." *Kansas Historical Quarterly* 19 (1951): 354–80.

Todorov, Tzvetan. *The Conquest of America: The Question of the Other.* Translated by Richard Howard. New York: Harper & Row, 1984.

Tree, Isabella. *The Ruling Passion of John Gould: A Biography of the British Audubon.* London: Grove Weidenfeld, 1991.

Turner, James. *Reckoning with the Beast: Animals, Pain, and Humanity in the Victorian Mind.* Baltimore: Johns Hopkins University Press, 1980.

Tyler, Ron. *Audubon's Great National Work: The Royal Octavo Edition of* The Birds of America. Austin, Tex.: W. Thomas Taylor, 1993.

Van Ravenswaay, Charles. *Drawn from Nature: The Botanical Art of Joseph Prestele and His Sons.* Washington: Smithsonian Institution Press, 1984.

Wainwright, Nicholas. *Philadelphia in the Age of Romantic Lithography.* Philadelphia: Historical Society of Pennsylvania, 1958.

Ward, Gerald W. R., ed. *The American Illustrated Book in the Nineteenth Century.* Winterthur, Del., and Charlottesville: Henry Francis du Pont Winterthur Museum, distributed by the University Press of Virginia, 1987.

Wayman, Dorothy G. *Edward Sylvester Morse: A Biography.* Cambridge: Harvard University Press, 1942.

Weber, David J. *Richard H. Kern: Expeditionary Artist in the Far Southwest, 1848–*

1853. Albuquerque: Published for the Amon Carter Museum by the University of New Mexico Press, 1985.

Weiss, Harry B., and Grace M. Zeigler. *Thomas Say: Early American Naturalist.* Springfield, Ill.: Charles C. Thomas, 1931.

Welker, Robert Henry. *Birds and Men: American Birds in Science, Art, Literature, and Conservation, 1800–1900.* Cambridge: MIT Press, 1955.

Wessen, Ernest J. "Jones' 'Nests and Eggs of the Birds of Ohio.'" *Papers of the Bibliographical Society of America* 47 (Third Quarter 1953): 218–30.

Winchester, Alice. *Versatile Yankee: The Art of Jonathan Fisher, 1768–1847.* Princeton: Pyne Press, 1973.

Zboray, Ronald J. *A Fictive People: Antebellum Economic Development and the American Reading Public.* New York: Oxford University Press, 1993.

Zboray, Ronald J., and Mary Saracino Zboray. "Books, Reading, and the World of Goods in Antebellum New England." *American Quarterly* 48 (1996): 587–622.

Zochert, Donald. "Science and the Common Man in Antebellum America." *Isis* 65 (1974): 448–73.

Index

Page numbers of illustrations are in italics.